W9-BUE-250

Also by James Lawrence Powell

Night Comes to the Cretaceous: Dinosaur Extinction and the Transformation of Modern Geology

Pathways to Leadership

James Lawrence Powell

MYSTERIES OF TERRA FIRMA

THE AGE AND EVOLUTION OF THE WORLD

THE FREE PRESS

NEW YORK LONDON TORONTO SYDNEY SINGAPORE

THE FREE PRESS
A Division of Simon & Schuster, Inc.
1230 Avenue of the Americas
New York, NY 10020

Copyright © 2001 by James Lawrence Powell
All rights reserved
including the right of reproduction
in whole or in part in any form.

THE FREE PRESS and colophon are trademarks
of Simon & Schuster, Inc.

Book design by Ellen R. Sasahara

Manufactured in the United States of America

10 9 8 7 6 5 4 3 2 1

Library of Congress Cataloging-in-Publication Data Is Available

ISBN 0-684-87282-X

All truth passes through three stages: First, it is ridiculed; Second, it is violently opposed; and Third, it is accepted as self-evident.

—Arthur Schopenhauer

Contents

Prologue

"YOU CAN OBSERVE A LOT BY WATCHING," said Yogi Berra. (Of course, he also said, "I really didn't say everything I said.") As the science of geology evolved during the nineteenth century, pioneer geologists lacked the tools to do more than observe the earth. But just "watching," in real time, hides most of Earth's mysteries. Careful study of rock layers did allow geologists to estimate the age of the earth in millions of years, but its true age, which measures in the billions, escaped them. Many careers spent observing failed to reveal that giant, continuously moving plates segment the earth's exterior. And in all of human history, no one has ever seen a meteorite fall from the sky and create a crater, providing no apparent reason to believe that meteorite impact has been an important geologic agent. All these discoveries had to await the arrival of the twentieth century and the invention and application of new tools for understanding Earth.

Of course, the revolutionary discoveries of the last century do not explain the meaning of our existence: that must be left to religion and philosophy. But the revolutionary findings of geology do place our planet and our species in time and space. They provide the ground truth with which any philosophy must reckon.

The Greeks believed that the moon and planets are made of shimmering crystal. But in 1609, Galileo Galilei (1564–1642), having improved the Dutch telescope, saw instead that the moon was a rocky world, like Earth. He observed Jupiter's moons circling like miniature planets and saw the countless stars of the Milky Way. Galileo's telescope revealed a universe of other worlds and a myriad of stars, each separated from its nearest neighbor by an immeasurable distance. Where, in Galileo's strange new universe, was our place?

In the eighteenth century, James Hutton (1726–1797) and his circle in Edinburgh concluded that catastrophes like the putative biblical Flood had not created the surface features of the earth, as most of his contemporaries believed. Rather, the same, slow, ordinary processes that anyone can observe at work today have sculpted the earth's surface. Hutton's circle embodied their concept in a slogan: The present is the key to the past. But if it is, earth history requires a long time—as Hutton put it, an abyss of time. To a universe filled with countless stars but simultaneously almost empty, the Huttonians added the immensity of geologic time.

On the shoulders of these disturbing and disorienting facts, the twentieth century built a revolution from three counterintuitive discoveries in the earth sciences. They transformed not only how we see ourselves and the earth, but how we view the origin of the solar system and the likelihood of intelligent life elsewhere. The three revolutions can be titled Time, Drift, and Chance.

The discovery of geologic time began with the English physicist William Thomson (1824–1907), knighted as Lord Kelvin, who used impeccable mathematics to prove that the earth can be no older than 100 million years. Later he lowered the limit to 20 million, which left biologists and geologists too little time for their theories. But Kelvin was wrong. Twentieth-century scientists found that the earth and the solar system are 4.5 billion years old and that the universe is 10 to 20 billion years old—time enough for two or three solar systems like ours to appear and die. Geologic time *is* an abyss: deep and not a little frightening.

The phrases *terra firma* and "solid as a rock" express our optimistic view of the reliability and permanence of the land beneath our feet. But we now know that the earth moves along great faults, and that, at depth, heat softens rocks until they melt. Geologists at the turn of the twentieth century believed "once an ocean, always an ocean; once a continent, always a continent." In this philosophy of permanence, continents did not move around, swapping places with ocean basins, but have always been where we now find them. If continents could move about willy-nilly, they might have been anywhere at any time. Then the present could not be the key to the past.

Yet, caught on thin, broad plates, continents do move. If all of earth history could be viewed in fast forward, the continents would perform a clumsy dance: now colliding to throw up a mountain range, now rifting apart to create a new ocean basin, now burrowing beneath one another and melting to create new crust. Plate tectonics not only taught us that *terra* is not so *firma,* it has had the ultimate practical value of constructing the very continental crust on which we stand and utterly depend. Without plate tectonics, the earth's surface would probably resemble the barren basaltic plains of Venus, or the sterile, forbidding face of Mars. But no one would be around to notice because without plate tectonics, intelligent life would not exist.

The third of the great twentieth-century discoveries—the role of random, violent meteorite impact in the solar system—drove home the lesson that the present is not the key to the past. The importance of impact has been easy to underestimate, for during the course of human history, not a single meteorite has struck the earth and created a crater. Indeed, until thousands of meteorites rained on a village in France in 1803, no one believed that rocks could fall from the sky. For a century and a half afterwards, scientists regarded meteorites as little more than curiosities.

The apparent tranquility and regularity of our solar system belie its history of incessant violence. Thanks to Copernicus, Newton, and the other astronomers, we know that the planets revolve in predictable orbits. The serene face of the moon shines on poets, lovers, and scientists alike, waxing and waning in what has become a metaphor of predictability. In spite of claims of early astronomers to have seen manbats on the moon and canals on Mars, when we look out at the solar system from Earth, little seems to happen. All appears calm; we are safe.

It took until 1994, and the collision of comet Shoemaker-Levy 9 with Jupiter, to remind us that all is not calm and that we are not safe. Our human time scale is too short to reveal the incessant nature and consequences of impact. The solar system was born in unimaginable violence and lives in violence. At its birth, bodies the size of planets collided at cosmic velocities. One giant impact knocked the earth atilt, blasted off a ring of material that became the moon, and sent both bodies careening. Another tremendous wallop blew away most of the surface of Mercury; still another flipped Uranus on its side. Countless objects disappeared into the inexorable maw of massive Jupiter. This Hadean cosmic bombardment melted the moon to its center and engulfed the earth in an ocean of magma hundreds of kilometers deep. Even after the planets cooled, orphan comets and asteroids continued to strike, as they do to this day and as they will do forever.

Impact is not only the most fundamental and energetic cosmic process, it is random, ruled solely by chance. If the tape of earth history were run again, an object of just the right size would not strike the earth at just the right angle and at just the right moment to tilt its axis by 24 degrees, giving us the seasons and the length of the day, and generating the moon. Next time, the meteorite that struck in the Yucatan 65 million years ago, if it had existed at all, would strike somewhere else, or miss the earth entirely. Then perhaps a dinosaur author would now be applauding the obvious cosmic destiny of the race to rule the earth not for a mere 160 million years, but for 225 million. But the one and only time the tape ran, fate smiled not on the dinosaurs and the other 70 percent of species that died with them, but on a small, hamster-sized creature scurrying beneath the dinosaurs' feet.

One day, just around the corner in geological time, the material contributions of geology—the ore and fuel deposits that the science has discovered—will be gone, or no longer needed. Then the three great, revolutionary discoveries of twentieth-century geology—Time, Drift, and Chance—will stand as the longest-lasting contributions of the field and among the most fundamental advances in human understanding. This book is the story of how scientists made those discoveries.

PART I

TIME

1

The Mill of Exquisite Workmanship

> If an elderly but distinguished scientist says that something is possible he is almost certainly right, but if he says that it is impossible he is very probably wrong.
>
> —ARTHUR C. CLARKE

A Gentle Confrontation

For geology, the twentieth century began on May 5, 1904, in a polite encounter between two physicists—one from the Victorian era and one whose work would launch twentieth-century science. On that day, geology started the long process of casting off the fetters of unfounded assumptions, uncritical deference to authority, and inability to measure and quantify.

The occasion was the meeting of the Royal Society of London, which had selected Ernest Rutherford (1871–1937), a New Zealand–born physicist and professor at McGill University in Montreal, to give its prestigious Bakerian Lecture. The audience of nearly eight hundred people included the cream of British science. They had come to hear the brilliant young scientist explain the results of his experiments in the new field of radioactivity, a name coined a few years earlier by the French physicist and Nobelist Marie Curie (1867–1934). Few in the audience could have fully understood the brave new world of transmutating atoms that Rutherford described. His science must have seemed closer to alchemy. For the most eminent scientist in the audience, Lord Kelvin, Rutherford's discoveries had special import: they cut the ground from under Kelvin's life's work. As Rutherford told it:

> I came into the room, which was half dark, and presently spotted Lord Kelvin in the audience and realized that I was in trouble at the last part of my speech dealing with the age of the earth, where my views conflicted with his. To my relief, Kelvin fell fast asleep, but as I came to the important point, I saw the old bird sit up, open an eye and cock a baleful glance at me! Then a sudden inspiration came, and I said Lord Kelvin had limited the age of the earth, provided no new source was discovered. That prophetic utterance refers to what we are considering tonight, radium! Behold, the old boy beamed upon me.

Kelvin began to study the shape and age of the earth as a teenager. He kept doggedly at it for more than sixty years, right up until his death three years after Rutherford's speech. With rare determination, Kelvin continually refined his estimate of the age of the earth. By the end of the century, he had reduced his original figure of 100 million years to 20 million, a span that most geologists thought too small to accommodate all that they had seen recorded in the rocks. Just when the geologists had enough, Rutherford came to their aid.

Close to the time of his speech to the Royal Society, Rutherford was walking the McGill campus. In his pocket he carried a small black object, a specimen of a uranium oxide mineral called pitchblende.

Meeting a colleague, Rutherford said, "Adams, how old is the earth supposed to be?" The answer came back at Kelvin's earlier figure of 100 million years. "I know," said Rutherford quietly, "that this piece of pitchblende is 700 hundred million years old."

Kelvin had calculated the age of the earth starting from the unquestioned assumption that both the sun and the earth had begun hot and had been cooling ever since. Rutherford did things differently. He took the piece of pitchblende into his laboratory, measured its radioactivity, and calculated its age directly. His figure of 700 million years turned out to be wrong; but considering that it was the first absolute age measurement and that he had made assumptions as well, it was not wrong by much.

Kelvin's assumptions would always remain unprovable. As the years passed and he adjusted his calculations again and again, his answer diverged further and further from the true age of the earth. In contrast, Rutherford's experiments steadily improved his assumptions, or converted them into facts. As a result, his answers became ever more accurate. Kelvin personified nineteenth-century science; Rutherford invented twentieth-century science.

By revealing that physicists could no longer explain natural phenomena entirely through the principles of mechanics and heat, radioactivity set the stage for the atomic physics and quantum mechanics of the twentieth century. But on that day in May 1904, the people gathered to hear Rutherford's lecture had no way of knowing what lay even a few years ahead. That did not stop Kelvin from pontificating. In a letter to his wife after his lecture, Rutherford wrote that he admired Lord Kelvin's confidence "in talking about a subject of which he has taken the trouble to learn so little." Kelvin even carried around, and was quick to show off, a small piece of radium that glowed in the dark. He never realized that the radioactive rays emanating from his amusing toy aimed straight for the heart of his assumptions.

Tonguestones

Though some Eastern religions were exceptions, not until the last two hundred years did people in the West think of the earth as having an age. The early Hindus believed in great cosmic cycles that repeated every 4,320,000 years; all the cycles taken altogether totaled 1,972,949,094 years. When Christian theologians eventually began to consider the age of the earth, they turned to the Bible. Archbishop James Ussher (1581–1656) invented the most famous and long-lasting biblical chronology. In contrast to the time scales based strictly on the Bible, Ussher also used astronomy and history. Eventually, he calculated that the earth was born on the "entrance of the night preceding the twenty-third day of October in the year of the Julian calendar 710." This translates to October 22, 4004 B.C. Thus, in Ussher's chronology, the earth is so young that its age measures in human generations. Science being too rudimentary to provide evidence to the contrary, well into the twentieth century Bibles still cited Ussher's date as the birth date of the earth.

Before the Enlightenment, the origin of fossils and sedimentary rocks was a mystery. Rather than being the remains of creatures that once lived, people thought that fossils were "Sports of Nature"—tricks planted to fool them or to test their faith. This view began to change with the discoveries and writings of a Danish anatomist and cleric named Niels Stensen, who Latinized his name to Nicolaus Steno (1638–1686).

Since antiquity, the island of Malta has produced the finest specimens of "tonguestones"—flattened, blade-shaped objects found embedded in rocks or lying on the surface. Some said the stones resembled a human tongue and possessed mysterious but useful properties, such as enhancing sexual prowess or controlling flatulence. Others thought them the perfect Sport of Nature, of no use whatsoever.

One day, sailors brought Steno the head of a Great White Shark to dissect. He noted that the shark's teeth, though smaller than the Maltese tonguestones, were in every other way identical to them. Steno deduced that the tonguestones were, or had once been, the teeth of sharks. But how had solid rock come to encase the tonguestones? In asking this seemingly innocent question, Steno laid his hand on the cloak of mystery that had always shrouded the age of the earth, and drew it back just enough to reveal the awful abyss beyond. The tonguestones disclosed, not an earth formed instantaneously and fully as we find it, but the passage of time. Sensing the chasm, unable to reconcile science and religion, Steno was eventually to recoil.

In considering how one solid could come to be inside of, or on top of, another, Steno concluded: "At the time when any given stratum was being formed all the matter resting upon it was fluid, and, therefore at the time when the lowest stratum was being formed, none of the upper strata existed." Such reasoning led him to deduce that a shark's tooth must be younger than the rock that encases it. Thus, rocks record not only divine creation, in which Steno believed, but the passage of time and the lives and deaths of creatures. But then biblical creation and earth history had to be reconciled. Steno could not do so, and he chose religion. He summed up his credo: "Beautiful is that which we see, more beautiful that which we know, but by far the most beautiful is that which we do not comprehend."

Hourglasses

Soon those who, unlike Steno, could set aside the dogma of the church, at least temporarily, began to try to calculate how much time might have passed since the earth formed. They used the principle of the hourglass. Over time, some quantity changes at a known or assumed rate, as when sand falls from the upper to the lower cone of an hourglass. If one knows how much sand is in the top and in the bottom of the hourglass, and the rate at which the sand falls through the constriction, one can calculate the length of time the process has been going on.

The accumulation of sedimentary rocks provided one of the first geological hourglasses. If one knows both the thickness of a sequence of rocks and the rate at which they accumulated, one can calculate how long the sequence took to form. Although the calculation is simple, it masks assumptions and difficulties that plague all the hourglass methods.

One obvious assumption is that one knows or can estimate the rate of accumulation; another is that the rate has been constant. The only non-circular way to estimate the past rate of sediment accumulation is to assume that today's rate equals the long-term average. But today's sedimentation rate could be higher or lower than the average, throwing off the result. (In fact, today's rate appears to be several times higher.)

Another assumption inherent in any hourglass method is that all of the sand that fell into the bottom cone of the hourglass is still there. We know that this is not true for sedimentary rocks, because they almost always contain gaps left where erosion has removed rocks. In that case, the total measured thickness of sedimentary rocks, and therefore the calculated age, will be low by some indeterminate amount. But at least the accumulation method gives a minimum age.

Instead of measuring how rapidly something builds up, one can measure how rapidly something else declines. If one knew how much erosion had lowered the land surface, and if one knew the rate, a simple calculation would reveal how long the erosion had been going on.

Unprovable assumptions underlie both the accumulation and the erosion hourglasses, yet early scientists had no other way of calculating ages. The temptation to fine-tune the hourglass of erosion led even Darwin into a calculation that he came to regret.

Telliamed and Buffon

The first to attempt to measure the age of the earth was a French diplomat and traveler, Benoît de Maillet (1656–1738). De Maillet assumed that a universal sea had once covered the earth but had since shrunk, stranding formerly coastal towns high above sea level. He estimated the rate of sea level decline at 3 feet in 1,000 years. At that rate, to perch a formerly seaside town at 6,000 feet would take 2 million years. But since the earth is obviously much older than its towns, de Maillet arbitrarily raised his estimate for its age to 2 billion years.

Well aware that such an immense figure would incur the wrath of the church, de Maillet presented his conclusions in the guise of a dialogue between a French missionary and an Eastern mystic named Telliamed (de Maillet spelled backwards). The manuscript remained unpublished until a decade after de Maillet's death. Then the priest to whom he entrusted his work moved the decimal point to the left. Since de Maillet had shifted the decimal point to the right, he could hardly have complained. But the correction did little good; Voltaire, among others, denounced de Maillet as a heretic.

Though his assumptions were wrong, de Maillet did show that, starting from observation and measurement rather than from the Bible, one could calculate an age for the earth. That age might be measured not in the scores of years by which a human lifetime is counted, nor even in the thousands of years of Ussher, but in millions and billions of years.

Georges-Louis Leclerc, comte de Buffon (1707–1788), left us a forty-four-volume treatise on natural history, and the aphorism, "Style is the man himself." Where de Maillet calculated, Buffon experimented. He postulated that a collision with a comet caused the sun to eject a long streak of matter that eventually condensed into the planets. The still molten protoplanets, spinning rapidly on their axes, in turn flung off smaller globules that became their moons. His theory led Buffon to an analogy between the once molten earth and a molten sphere of metal.

Buffon built iron spheres of different sizes, heated them up, measured how long it took them to cool, and extrapolated to a ball of iron the size of the earth. He got ages as great as 3 million years, but lowered his published result to 75,000 years. Even that proved too much for the theological faculty of the Sorbonne, who instructed Buffon to reduce his estimate.

"You Can't Win"; "Eventually, You Lose"

Over the eighteenth and nineteenth centuries, descriptive geology slowly progressed. Steno's principle of superposition and the pioneer English geologist William Smith's discovery of index fossils allowed geologists to construct their standard geologic column—the ideal section of rock in which every known formation is shown in its correct thickness and position. The thickness of Paleozoic and younger sedimentary rocks (see the geologic time scale in Figure 1.1) clearly totaled tens of thousands of feet. This was far more than could have accumulated in the few thousand years to which biblical chronology limited earth history. Thus, by the middle of the nineteenth century, though geologists suspected that the age of their planet must measure in millions of years, they had no way of knowing just how many millions.

With the publication of Sir Charles Lyell's *Principles of Geology* in 1830–33, geologists got more millions than they bargained for. Lyell rejected the catastrophism of floods, impacting meteorites, and violent upheavals. He argued persuasively that it is unscientific to claim as an agent of geologic change any process that we cannot observe operating today. A Cambridge professor, William Whewell (1794–1866), gave Lyell's theory the unwieldy name "uniformitarianism."

Lyell claimed not only that natural laws and processes are constant, but that erosion, deposition, and the like have always operated at the

Eon	Era/subera		Period/subperiod		Epoch	Estimated age (Millions of Years)
			Quaternary		Holocene	.01
Phanerozoic	Cenozoic				Pleistocene	1.6
		Tertiary	Neogene		Pliocene	5.1
					Miocene	24
			Paleogene		Oligocene	38
					Eocene	55
					Paleocene	65
	Mesozoic		Cretaceous			144
			Jurassic			213
			Triassic			248
	Paleozoic		Permian			286
			Carboniferous	Pennsylvanian		320
				Mississippian		360
			Devonian			408
			Silurian			438
			Ordovician			505
			Cambrian			570
Proterozoic	Late					900
	Middle					1,600
	Early					2,500
Archean	Late					3,000
	Middle					3,500
	Early					4,000
Priscoan						4,550

Figure 1.1 The Geologic Time Scale (after Dalrymple, 1991).

same rate. He asserted that constancy in one part of the globe offsets change in another, leaving the whole the same. If the processes that shape the earth, and the rate at which they operate, are constant, then over the long run the appearance of the earth must also be constant. Lyell's philosophy not only required immense amounts of time, it offered no way of limiting time. For Kelvin, Lyell's unlimited draft on the bank of time violated the laws of nature.

The first law of thermodynamics holds that in any process, energy in the form of heat and work is conserved. As science students once liked to joke, the first law states, "You can't win"—one cannot get more energy out than one puts in. The second law, jointly discovered by Kelvin and the German mathematical physicist Rudolf Clausius, states that every thermodynamic process flows in only one direction and afterwards cannot be returned to its original state. Without the intervention of an external device, heat will not flow from a lower to a higher temperature. The message of the second law is, "Eventually, you lose."

In describing geologic time as infinite and the earth as unchanging, Lyell claimed that the earth is a perpetual motion machine, one that can not only win the energy battle but go on doing so forever. But the first and second laws prove that a perpetual motion machine is impossible. Kelvin carried the battle to the geologists, charging that "a great mistake has been made—that British popular geology at the present time is in direct opposition to the principles of Natural Philosophy." For "Natural Philosophy," read physics. Geology would have to bow to physics and Kelvin would see that it did.

Colossus

Kelvin got an early start in thinking about the earth. In 1840 as a sixteen-year-old university student, he wrote an essay called "On the Figure of the Earth." Though the paper won the University at Glasgow medal, Kelvin never published it. Perhaps he was never satisfied—for sixty-seven years he continued to refine his essay, giving it a last polishing only two months before his death. Stephen Jay Gould has written that Charles Lyell "doth bestride my world of work like a colossus"; but for half a century, Kelvin bestrode the entire world of science, dominating biologists and geologists alike.

By limiting the age of the earth and the sun, Kelvin would put right the great mistake of geology. Whatever the limit turned out to be, it would be less than Lyell's infinity of time and would return geology to concordance with the second law. Since Kelvin could not measure the age of the earth and the sun directly, he had to estimate their ages from theory, which required that he have a model. He began with the nebular hypothesis for the origin of the solar system, in which he had excellent company.

Newton's laws of motion and gravitation revealed the fundamental "glue" that holds the clockworklike apparatus of the solar system together, but they did not explain how the system began. In 1755, Immanuel Kant (1724–1804) postulated that the solar system started as a cloud, or nebula, of particles. Newtonian gravitational attraction drew the particles into clumps, which, growing larger by accretion, eventually became the planets. Kant, as a philosopher, did not realize that his model failed to explain why the planets revolve around the sun in the same direction and in the same plane.

Four decades later, the great French polymath Pierre-Simon Laplace (1749–1827) improved Kant's model. Laplace's theory began with the sun cooling and contracting. Since angular momentum is maintained, the shrinking, whirling sun had to spin faster, increasing the centrifugal force on its outer regions. At some point, the centrifugal force that tended to fling material outward just exceeded the gravitational force that tended to draw it in, pinching off blobs of solar material and launching them as planets. The planets recapitulated the process to create their moons. By Kelvin's day, the nebular hypothesis had become the starting point for any system of cosmogony. The theory was so deeply engrained that nearly everyone, Kelvin included, forgot that it was only a model, and one not subject to experiment or proof.

To derive the age of the earth starting with the primordial nebula, scientists needed a mathematical construct. The French mathematician Joseph Fourier (1768–1830) had worked out the theory of heat conduction in a solid body. He showed that by knowing how rapidly temperature increases with depth inside the earth, one can calculate how long it would have taken a body the size of the earth to cool to its present surface temperature. Fourier calculated an age so far out of line with current views that he did not even bother to write it down: 200 million years. At age seventeen, the precocious Kelvin had already learned Fourier's theory well enough to correct a mistake in its use made by a professor at the University of Edinburgh. Years later, when he wished to correct the mistake of geology, Kelvin again turned to Fourier.

At first, the idea that as meteorites fall into the sun, Icarus-like, the gravitational energy they lose converts to heat, attracted Kelvin. But calculations showed that the energy released by falling meteorites was inadequate to explain the sun's abundant heat and light. In the absence of any other source, the sun's energy must be that left over from its birth. The second law then requires that the sun be running down. If the sun's energy is waning, it must have been hotter in the past and will one day be cooler, so cool as to be unable to warm and illuminate the earth. No sooner had Kelvin helped to discover the second law of thermodynamics (at the age of only twenty-eight) than he used this reasoning to set out the fundamental conviction that was to guide his work for the rest of his life:

> Within a finite period of time past the earth must have been, and within a finite period of time to come the earth must again be, unfit for habitation of man as at present constituted, unless operations have been, or are to be performed, which are impossible under the laws to which the known operations going on at present in the material world are subject.

In a famous 1862 article in the popular *Macmillan's Magazine,* Kelvin described the results of his calculations of the age of the sun:

> It seems, therefore, on the whole most probable that the sun has not illuminated the earth for 100,000,000 years, and almost certain that he has not done so for 500,000,000 years. As for the future, we may say, with equal certainty, that inhabitants of the earth cannot continue to enjoy the light and heat essential to their life, for many millions of years longer, unless sources now unknown to us are prepared in the great storehouse of creation.

To Kelvin, whether the sun was 100 million years or 500 million years old mattered little. The earth could be no older. By limiting the age of the sun, Kelvin had refuted Lyell.

In the last several words of each of the two quotations, Kelvin wisely ceded that he and his contemporaries did not yet know everything. He explicitly recognized that future discoveries of laws, operations, and sources of energy might invalidate his assumptions and conclusions. The prospect of unknown sources of light and heat was especially prophetic, for a "storehouse of creation" unknown to Kelvin did exist: radioactivity. But its discovery and significance lay four decades in the future. Never once, not even after the discovery of radioactivity, did the obstinate Kelvin heed his own wise counsel.

In 1862, Kelvin turned from the sun to the earth and to Fourier's theory. His first calculation gave the same age for the earth as Fourier: 100–200 million years. In 1865, Kelvin wrote a paper entitled "The 'Doctrine of Uniformity' in Geology Briefly Refuted," as though, in one "brief" paper, physics could do away with the underlying premise of geology and have geologists willingly submit. As the historian of science Stephen Brush has observed, "Apparently Kelvin was infected by the common fallacy that an established scientific theory can be immediately overthrown by citing a single devastating argument against it." Nothing could be further from the truth.

Throughout his long life Kelvin returned repeatedly to the age of the earth, incorporating the latest heat flow and other new data. Though his method remained the same, his answer fell with "harmonic regularity": "By 1868 . . . 100 million years. In 1876 . . . 50 . . . and in 1881 . . . 20 to 50 million years. Finally, in 1897 . . . 24 million years." Biologists and geologists, who required ever-expanding periods for their theories, must have feared that Kelvin was going to leave them no time at all.

Lord Kelvin's Opponents

From the first, Kelvin's opponents provided easy targets. Lyell made no effort to estimate the earth's age and indeed did not seem to conceive of the planet as having an age. He claimed that although it continually loses heat to space, a thermoelectric engine in the earth's interior allows it to remain at the same temperature indefinitely. To give him credit, Lyell came up with this idea before Clausius and Kelvin discovered the second law. But like Kelvin and others we will meet, when he had the chance to change his mind, Lyell never took it. Against the adamantine second law of thermodynamics, Lyell offered a perpetual motion machine, a prospect as unlikely as his view that extinction is not forever and that, one day, the wings of the pterodactyl may again flap over a forest primeval.

In the 1868 address to the Geological Society of Glasgow that began his attack on the great mistake of British geology, Kelvin directed his argument not at the living and formidable Lyell, but at the long-dead John Playfair, interpreter of the inarticulate James Hutton. As an example of geological reasoning, Kelvin chose Playfair's sixty-six-year-old statement: God "has not permitted in his works any symptoms of infancy or of old age, or any sign by which we may estimate either their future or their past duration." Playfair appeared to be saying that time is limitless and any attempt to measure it is pointless. To the discoverer of the second law, this was scientific heresy.

Those two champions of evolution, Charles Darwin (1809–1882) and Thomas Huxley (1825–1895), fared no better. Lyell's *Principles* had so impressed Darwin that he took a copy aboard HMS *Beagle* and later wrote, "He who can read Sir Charles Lyell's grand work and yet does not admit how incomprehensibly vast have been the past periods of time, may at once close this volume." Lyell provided Darwin with the time he required.

Only once did Darwin attempt to calculate a geologic age, and there he fell into a trap. He chose to estimate how long it had taken to erode a large dome in Kent, southern England, called the Weald. Darwin estimated that "for a cliff 500 feet in height, a denudation of one inch per century for the whole length would be an ample allowance." He then concluded, "At this rate, the denudation of the Weald must have required 306,662,400 years; or say three hundred million years." Everyone scoffed at the notion that one valley could be several times older than Kelvin's earth. Before long, Darwin was referring to "those confounded millions of years."

Thomas Huxley, Darwin's fierce and stubborn "Bulldog," had by all accounts routed the "shallowly-eloquent" Bishop Wilberforce in their famous debate of 1860. Wilberforce patronized Huxley by asking whether it was through his father or his mother that he had descended from an ape. Huxley countered that he would not be ashamed of an ancestor who was an ape but would be of one who used gifts of eloquence in the service of falsehoods. Now, nine years later, Huxley's opponent was not the condescending cleric, but Lord Kelvin, the man acknowledged by even his opponents as the leading British scientist.

Huxley and Kelvin did not confront each other directly. In his 1868 address, Kelvin fired the first shot, calling out the error that geologists had made, and were continuing to make, in claiming that the age of the earth had no limit. The latest edition of Lyell's *Principles* had just appeared; again Lyell ignored what Kelvin regarded as fundamental laws of nature. Now Lyell even admitted that he called for a perpetual motion machine.

As president of the Geological Society of London, Huxley may have had little choice but to answer Kelvin. In 1869, Huxley tried to do so, but found himself in the same weak position in which many geologists over the next one hundred years were to find themselves: unable to refute an apparently superior quantitative argument from a physicist. Huxley could not understand, much less counter, Kelvin's mathematics, so he had to fall back on "mother-wit" and his considerable rhetorical skills. Huxley pounced on Kelvin's selection of the long-dead Hutton and Playfair as his targets, arguing that geologists had long since modified the overly rigid uniformitarianism of the two founding fathers. "Catastrophes may be part and parcel of uniformity," Huxley claimed.

Huxley's most telling point, and the most quoted statement from his speech, came after a long attack on Kelvin's many assumptions. He elegantly summed up what today we often put more crudely: "Mathematics may be compared to a mill of exquisite workmanship, which grinds you stuff of any degree of fineness; but, nevertheless, what you get out depends upon what you put in." But Kelvin would not let Huxley off with an appeal to mere mother-wit: "The very root of the evil to which I object is that so many geologists are contented to regard the general principles of natural philosophy, and their application to terrestrial physics, as matters quite foreign to their ordinary pursuits." Since his opponents could not refute his impeccable mathematics, and since they could offer no better alternative, Kelvin prevailed. Huxley's address was the last time for several decades that anyone would challenge Kelvin. He had shifted the ground of debate about the earth. No longer could Lyell's limitless time be endorsed, nor his unchanging earth.

Swept Away

Geologists set about trying to measure the length of geologic time by their own methods, part of a general effort during the second half of the nineteenth century to base their science more on measurement than description. But after laboriously calculating the age of the earth by one of their hourglass methods, geologists required an external reference to certify their result as reasonable. Yet only one reference existed: the apparently exquisite mathematical calculation of the leading scientist of the day, Lord Kelvin, who had determined the age of the earth to be 100 million years. To use a geological method and reach the same conclusion as Kelvin not only validated one's scientific acumen, it confirmed the stature of the discipline of geology. The combination proved irresistible.

Though not many occupied themselves with the age of the earth, those who did "produced an amazing variety of methods and an even greater homogeneity of results." No matter what assumptions and approaches they used, the hourglass calculators wound up agreeing with Kelvin. Two examples illustrate the fragility of their assumptions and the malleability of their results.

T. Mellard Reade (1832–1909) tried the hourglass of erosion, employed to his regret by Darwin. Reade assumed that the crust of the earth under the seafloor has the same composition and thickness as the crust under the continents. He assumed that both the surface area of the earth undergoing erosion and the rate of erosion have been constant. His calculations, which appeared in the 1870s, produced an age of 600 million years, which Reade viewed as a minimum and which he initially defended against Kelvin's much lower limit.

When the Challenger oceanographic expedition of the 1870s, sponsored by the British Admiralty and the Royal Society, found that the crust of the seafloor is not sedimentary, but is mainly basalt, Reade had to adjust his calculation. He also decided, arbitrarily, to assume that the area undergoing erosion is the same as the area that receives the eroded sediments. Next, he introduced a correction to recognize that sedimentary material is recycled: sedimentary rock erodes into sediment that hardens into sedimentary rock that erodes again, and so on. When the adjustments were over, Reade's result had shrunk to 95 million years, allowing him to say that the earth's age is "somewhere between 100 and 600 million years," thus preserving his original figure while still allowing Kelvin's.

Samuel Haughton (1821–1897), professor of geology at Trinity College, Dublin, made the most bizarre and revealing series of calculations. He first concluded that all the time before the Tertiary amounted to over 2 billion years, far above Kelvin's limit. Haughton then invented a peculiar and incomprehensible method of using fossils to estimate the rate of decline of the earth's temperature. This led him to just the opposite conclusion: The time since the Miocene epoch is greater than all the time that preceded it—the Paleozoic era, the Mesozoic era, and all the pre-Miocene part of the Cenozoic era put together (cf. Figure 1.1). Haughton maintained this position even though it is obvious that far more sedimentary rock predates the Miocene than postdates it. He then went on to assume a much larger area over which sedimentation takes place, and a much greater thickness of accumulated sedimentary rock, than did others. His assumptions, uncorrected, gave an age of about 1.5 billion years. Following in the footsteps of Telliamed and his priest, as the final step in his calculation, Haughton capriciously divided by ten.

Haughton's 150-million-year result entered the literature and remained, his arbitrary methodology forgotten, as yet another confirmation of Kelvin's infallibility. Geologists accepted any result that came close to Kelvin's, no matter how contrived, as further corroboration that geology could stand beside physics. Galileo, Steno, Telliamed, and Buffon bowed to God; the Victorian geologists bowed to Lord Kelvin.

But the geologists were not the only ones to pay homage. George

Howard Darwin (1845–1912), second son of Charles, was a protégé of Kelvin and later a professor at Cambridge. He served as president of the Royal Astronomical Society and president of the British Association. Like his contemporaries, George Darwin subscribed to the nebular hypothesis, according to which the moon had started out close to the earth and has since gradually receded to its present location. One skilled at mathematics might be able to work backwards and find out when the moon had begun its retreat from the proto-earth—the birthday of the solar system.

Darwin calculated that for the earth and moon to achieve their present separation would have taken 56 million years. He cautioned, "The actual period, of course, must have been much greater," saying that his calculation "is only a wild speculation, incapable of verification." Unfortunately, Darwin could not resist noting that his result fell within the range established by Lord Kelvin. Like other estimates of Earth's age in the second half of the eighteenth century, George Darwin's estimate of 56 million years entered the literature as yet another confirmation, and by a rigorous and independent method, of Kelvin's accuracy. No one remembered Darwin's caution.

As certainty replaced mere confidence, Kelvin and his followers squeezed the stratigraphers even more. One sycophant proclaimed in 1876 that it is "utterly impossible that more than ten or fifteen million years can be granted." Geologists could scarcely fail to feel the pinch, nor resent the tone, according to which it was neither God nor Nature, but Kelvin, who granted geologic time.

Just as British geologists began to muster the courage to stand up to Kelvin, support for him came from a new source, the founding director of the U.S. Geological Survey, Clarence King (1842–1901). King had advocated the use of quantitative methods in geology and had the private funds to set up his own laboratory and practice what he preached.

Kelvin had assumed that the earth had been initially molten and at a temperature of 3,900 degrees Celsius. King started there, but having more recent information about the melting point of rocks and the distribution of temperature within the earth, he could extend and refine Kelvin's calculations. King settled on 24 million years as the age of the earth.

In an 1897 address, his last on the subject that had preoccupied him throughout his long scientific life, Lord Kelvin pronounced himself in agreement with King: 24 million years was just right. In a tone of triumphant certainty, Kelvin proclaimed that King had reconciled the age of the earth, as determined from solar heat, with that calculated from terrestrial heat. Kelvin announced that the reconciliation "suffices to sweep away the whole system of geological and biological speculation demanding an 'inconceivably' great vista of past time, or even a few thousand million years, for the history of life on earth." But new advances would sweep Kelvin's conclusions away.

The Bank of Time

Over the course of the second half of the nineteenth century, geologists had gradually accommodated themselves to Kelvin's 100-million-year age for the earth. They found ingenious, and sometimes disingenuous, ways of confirming it. Huxley's appeal to "mother-wit" exposed the inadequate arsenal of biologists and geologists in the face of Kelvin's weaponry of exquisite calculations. They had no choice but to capitulate. The first leading British geologist publicly to accept Kelvin's limited time scale was Sir Archibald Geikie (1835–1924), director of the Geological Survey of Scotland and later director general of the Geological Survey of the United Kingdom. In paper that appeared in 1871, just after the Huxley-Kelvin debate, Geikie endorsed Lord Kelvin's 100 million years. "We have been drawing recklessly upon a bank in which it appears there are no further funds at our disposal," Geikie wrote. "It is well, therefore, to find that our demands are really unnecessary."

But it was Kelvin's demands that eventually proved too much. No sooner would the Victorian geologists accede to his latest, always lower, limit than Kelvin would lower it again. Opposition grew, and as century's end approached, geological opinion began to turn. In 1892, Geikie reversed himself, declaring that "some assumption has been left out of sight." Expressing his frustration with the overbearing Kelvin, Geikie fumed, "It is difficult satisfactorily to carry on a discussion in which your opponent entirely ignores your arguments, while you have given the fullest attention to his."

Other geologists joined Geikie in stressing Kelvin's unfounded assumptions. One noted that "The utmost any physicist is warranted in affirming is that it is impossible for him to conceive of any other source [of the sun's energy]. His inability, however, to conceive of another source cannot be accepted as proof that there is no other source." In other words, ignorance is no foundation for certainty.

The criticism from closest to home came from Kelvin's former assistant and partner, John Perry (1850–1920). As Kelvin relentlessly reduced the possible duration of geologic time while escalating his professed certainty, Perry came to believe that it was his "duty to question Lord Kelvin's conditions." He showed that only a slight change in Kelvin's assumptions could produce quite a different answer for the age of the earth. In a sincere but naive statement, Perry wrote to a supporter of Lord Kelvin that "as soon as one shows that there are possible conditions as to the internal state of the earth," Kelvin's case was undone. Kelvin met Perry partway by conceding that based on terrestrial heat alone, the age of the earth could be as great as 4,000 million years. But still the sun's heat limited the age of the solar system, and therefore the age of the earth, to a few score million years. Only two years later, Kelvin endorsed King's 24 million figure. But Kelvin's mill of exquisite workmanship was about to find sand in its gears—sand from the ultimate hourglass.

When the American geologist Thomas Chrowder Chamberlin

(1843–1928) read the text of Kelvin's speech, he took offense at Kelvin's authoritarian language: "half an hour after solidification" and "certain truth." Chamberlin was in a strong position to challenge Kelvin. He had been president of the University of Wisconsin and was chair of the geology department at the University of Chicago. Years earlier, Chamberlin had pointed out the disadvantages of rigid adherence to a single explanation. He advocated instead "multiple working hypotheses," in which minds are kept open to different interpretations for as long as possible.

Together with Forest Ray Moulton (1872–1952) a young astrophysicist, Chamberlin had begun to work out an alternative to the nebular hypothesis of Kant and Laplace. He and Moulton thought that instead of a hot nebula flinging off the planets, they may have accumulated from cold, tiny, "infinitesimal planets," or "planetesimals."

Chamberlin and Moulton challenged the premise on which Kelvin based all his calculations. That the earth had once been completely molten, Kelvin had described as a "very sure assumption." Chamberlin "beg[ged] leave to challenge," arguing that the earth may have formed slowly from cold planetesimals and therefore may never have melted entirely. Consistent with his philosophy, Chamberlin did not claim that he was right and Kelvin wrong, only that either could be right.

Chamberlin also attacked Kelvin's age for the sun, arguing that no one really knew the source of the sun's energy. His words were prophetic:

> Is present knowledge relative to the behavior of matter under such extraordinary conditions as obtain in the interior of the sun sufficiently exhaustive to warrant the assertion that no unrecognized sources of heat reside there? What the internal constitution of the atoms may be is yet an open question. It is not improbable that they are complex organizations and the seats of enormous energies. Certainly no careful chemist would

affirm either that the atoms are really elementary or that there may not be locked up in them energies of the first order of magnitude.

With great prescience, Chamberlin had set the stage for the ideal hourglass. But first, geologists were to have a last go with their methods.

The Salt Clock

In 1715, Edmund Halley (1656–1742), discoverer of the eponymous comet, proposed that an hourglass of salt would reveal the age of the oceans. If one knew the rate at which streams and rivers dissolve salt from rocks and deliver it to the oceans, and if one knew the total amount of salt in the sea, one could calculate how long the process had been going on, which would equate to the age of the oceans. In Halley's day, no one knew either, so his idea had to be rediscovered in the 1870s. John Joly (1857–1933), professor of geology and mineralogy at Trinity College, Dublin, made the method his own (using sodium rather than sodium chloride). By the end of the nineteenth century, scientists knew that the oceans contain about 1.42×10^{16} metric tons of sodium and that streams deliver about 1.43×10^{18} metric tons of sodium per year. Thus, the age of the oceans, which Joly equated with the age of the earth, is 99 million years. Joly attempted to correct for the presence of original sodium in the primordial oceans, which lowered his estimate to 89 million years. This number Joly defended for the rest of his life, thirty-three years into the new century, and long after it had become clear that radioactivity had obviated the Kelvin time scale.

As enthrallment with Kelvin began to wane, geologists recognized that their methods, although they gave older and therefore more acceptable ages than Kelvin's, included so many assumptions that they were bound to be inaccurate. One scientist listed four possible sources of error in the stratigraphic clock:

- The total accumulation of sediment occurs in no one spot;
- Sediment accumulates at greatly different rates;
- The duration of time represented by unconformities (gaps) is unknown;
- Sedimentary particles are recycled through erosion, deposition, and erosion again.

An even more fundamental assumption was that the current rates of geologic processes are close to the average rate over all of geologic time, and that those rates have not varied. Finally, since at the time only relative rock ages from fossils could be established, and since Precambrian rocks contain no macrofossils, geologists of the nineteenth century had no way of knowing that Precambrian time makes up almost 90 percent of geologic time. That alone was to reduce their calculated ages to only a fraction of the true age of the earth.

Joly's salt clock contained an additional, unsuspected flaw. When we repeat his calculation using modern data for the sodium content of rivers and oceans, we obtain an "age" of about 68 million years, not far below Joly's later calculations. But we know that the earth and the oceans are vastly older than that. If Joly's method is not wrong because of the poor quality of data put into the equation, why is it wrong? Unknown to Joly, geological processes have carried some of the sand in the bottom of his hourglass back to the top. This error caused him to derive neither the age of the earth nor the age of the oceans, but of something else entirely.

Joly and his contemporaries incorrectly assumed that all the salt delivered to the oceans remains dissolved in seawater. In the 1960s, scientists discovered that as erosion adds chemical elements to the oceans, and deposition removes them, seawater reaches chemical equilibrium.

After that, the more of an element added, the more removed, leaving the average concentration everywhere the same. In particular, much of the sodium that reaches the oceans evaporates or joins sedimentary rocks, which are later exposed to erosion and whose sodium returns to the sea. It turns out that Joly had measured, not the age of the oceans, but the average length of time that a sodium atom, delivered to the sea, remains dissolved.

As the nineteenth century closed, the era of flawed methods began to close as well. An hourglass of inexorable precision was about to replace all the previous hourglass methods. Elaborate calculations of the age of the earth based on untestable assumptions were to go the way of the gas lamps that lit British streets. The settling of scientific disputes by rhetoric was likewise to decline, a victory for reason if a defeat for eloquence. The twentieth century awaited with methods for measuring the age of the earth that were to brook little debate. All the efforts of nineteenth-century scientists to measure and limit the age of the earth, Lord Kelvin's included, came to naught. Sadly, he and his contemporaries had not been able even to prepare the ground.

2

Strange Rays

An important scientific innovation rarely makes its way by gradually winning over and converting its opponents. . . . What does happen is that its opponents gradually die out and that the growing generation is familiar with the idea from the beginning.

—MAX PLANCK

EVEN AS LORD KELVIN proclaimed the earth to be only 24 million years old, research that would prove him wrong had begun. The new findings would launch geology and physics into the twentieth century, and eventually, threaten the survival of our species. As often happens, the breakthrough began in serendipity.

Before the turn of the century, the French physicist Antoine-Henri Becquerel (1852–1908) studied minerals that glow when exposed to light and that continue to do so even after the light is turned off, a property called phosphorescence. Becquerel was trying to determine the relationship between phosphorescence and the X rays that William Roentgen had just discovered. Suspecting that sunlight triggers phosphorescence, Becquerel wrapped a photographic plate in black paper so that no light could reach it, placed a specimen of a phosphorescent uranium mineral on top of the wrapped plate, and sat the whole thing in the sunlight. Sure enough, when he developed the plate, he found that it had captured the image of the uranium crystal. But, for a few days, the Paris sun failed to shine. Becquerel, still believing that sunlight was key, suspended his experiment and stored the crystal atop a photographic plate, but inside a drawer where no light could reach it. When he later developed the plate, expecting to find only a faint impression due to the residual phosphorescence of the mineral, he found as sharp an image as ever. Becquerel realized that sunlight was not essential after all; instead, the image came from a new kind of ray emitted by the crystal itself.

Becquerel suggested that his student, Marie Curie, take up the study of the new rays. Thus was launched one of the most productive careers in science. Like any graduate student then and now, Curie needed to choose a Ph.D. thesis topic. She decided to investigate whether any other substances emitted the uranium rays, which she and Becquerel had called "radioactivity." Quickly, she (and independently a German scientist) found that a mineral containing the element thorium also gave off radioactivity. She then discovered that pitchblende emitted more radioactivity than the uranium contained within the pitchblende. She and her husband, Pierre (1859–1906), concluded that pitchblende must contain unrecognized sources of radioactivity. They discovered two; one they named one polonium, after her native Poland; the other they named radium.

In 1903, Marie Curie, her husband, and Becquerel shared the Nobel Prize in Physics. In 1911, she won the Nobel Prize in Chemistry, making her the only person to win the prize in both physics and chemistry. In 1934, she died from leukemia contracted during her years of exposure to radiation.

Rutherford

The new phenomena of X rays and radioactivity attracted the interest of many physicists. One was a young New Zealander, Ernest Rutherford, whose career personifies the benefits to society of the educational scholarship. Rutherford's first award allowed him to attend Nelson College on the South Island of New Zealand. Having excelled at Nelson, he next received a scholarship to Canterbury College of the University of New Zealand. He did so well there that he won a scholarship to Cambridge University, where his intellect and gift as an experimentalist soon made him stand out.

Rutherford and his thesis professor, J. J. Thomson (no relation to Kelvin), discovered that when X rays travel through a gas, they engender a host of charged atoms, called ions. Rutherford wondered whether Becquerel's new uranium rays would also ionize air. He found that they did, but in the process learned that the uranium rays are not X rays. Becquerel had already shown that when the uranium rays pass through a magnetic field, some veer off to one side, indicating that they have a negative charge, while others veer to the opposite side, showing that they have a positive charge. Rutherford named the negative rays "beta" and the positive rays "alpha," the names by which we know them today. The beta rays turned out to be the electrons that J. J. Thomson had discovered. Rutherford was intent on identifying the alphas, a search that took him a quarter of a century but that bore a rich scientific reward.

In 1898, Rutherford moved to McGill University in Montreal, where he continued to demonstrate his great gift as an experimentalist. One of his young colleagues found that he could blow the radioactivity emitted by thorium about in the air—the thorium rays were a gas. Rutherford drew the gaseous rays into a tube and found that every 54.5 seconds, their activity declined by half. No matter how much of the thorium gas he started with, in 54.5 seconds the amount of radioactivity it gave off fell by exactly one half. In the next 54.5 seconds, the activity would decline by half again, and so on. Rutherford found that other substances also lost half their activity in a given amount of time, but that the time differed for each. Thus was born the concept of half-life. (The gas produced by the decaying thorium was radon 220.)

Another component of thorium, called thorium X, decayed with a half-life of 3.6 days, but built up again in the parent thorium at the same rate. Here was a strange new world: substances decay so as to lose exactly half their original activity in a fixed amount of time, but in so doing they transmute themselves into other substances. These new products then die away, but with a different half-life. Far from being eternal and immutable, an atom of one sort, with no warning, may disappear, and another of an entirely different sort may arise in its place.

In 1902, Rutherford and Frederick Soddy published a classic paper that defined and explained radioactive decay and growth. Radioactive decay obeys the law of probability. If one could observe a particular atom, say of radon 220 with its half-life of 54.5 seconds, one could not predict when that atom would decay. It might happen within the first 5 milliseconds of observation, or take hours, days, weeks, or years. One can speak only of the probability that the atom will decay within a certain length of time. But if the number of atoms observed is large enough, the laws of probability dictate that in one half-life, exactly one half that number will decay.

Take the familiar example of a tossed coin, with its 50–50 chance of showing heads. A coin tossed only a few times may well give heads each time, or tails each time, or any combination—the results are unpredictable. But a coin tossed more and more often gives a percentage of heads that approaches and finally reaches 50 percent to whatever number of significant figures one (or one's computer) has the stamina to produce. Toss a perfectly honest coin 10^{23} times, for example, and the number of heads will be 50 percent to many significant figures. Ten raised to the twenty-third power is approximately the number of molecules in only one gram-mole of any chemical element (Avogadro's number: 6.02×10^{23}). Even in the smallest laboratory sample, so many atoms are present that overall they obey the law of radioactive decay with complete fealty.

Other familiar processes also obey a probability law. Take, for example, the number of births in a population. If one knows the birth rate (number of births annually as a percent of population) and the size of the population, one can predict how many children will be born in one year. The larger the population and the better known the birth rate, the more accurate will be the prediction. But no one can foretell whether, more than nine months in the future, a given couple will have a child. That is a matter of probability. To use the opposite, macabre example, imagine that all human beings become instantaneously sterile so that no child is ever again born. Then the world's population would begin to decline with some half-life and continue to do so until, more than a century into the future, our species would be extinct. But between the final two survivors, there would be no statistical way to predict who would die first.

In the early years of the twentieth century, astounding scientific discoveries piled atop each other. In 1903, Rutherford found that atoms expel alpha particles at the fantastic speed of 24,000 kilometers per second. Pierre Curie discovered that radium emits heat: one gram gives off about 100 calories per hour—enough to raise the temperature of a gram of water from the freezing to the boiling point. Rutherford went on to show that the amount of radioactive heat produced is directly proportional to the number of alphas emitted. He concluded, as Chamberlin had suggested with great foresight in his rebuttal to Kelvin, that atoms are indeed the sites of enormous stores of energy.

Rutherford was almost certain that alpha particles are atoms of the element helium. Having measured the velocity of the alphas, and guessing their mass, he calculated their energy using the formula kinetic energy equals one-half the mass of a particle times its velocity squared.

Even though they are small, the alpha particles travel at such speeds that they contain a great deal of energy. As the tiny but energetic alpha particles fly outward, the crystal lattices of the material through which they pass eventually absorb the particles. Since energy is conserved, the kinetic energy of the alphas converts to heat, the heat that Pierre Curie discovered.

In his cautions about "operations . . . which are impossible under the laws to which the known operations going on at present in the material world are subject," and "sources now unknown to us . . . prepared in the great storehouse of creation," Kelvin seemed almost to anticipate the discovery of radioactivity. But after its discovery, and ever after, Kelvin never acknowledged that the existence of radioactive heat required him to change his assumptions. At various times, he accepted radioactivity, denied it, and finally, ignored it.

At the meeting of the Royal Society in 1904, where the young Rutherford faced the somnolent Kelvin, only a few felt Kelvin's house of cards begin to teeter. If, as Rutherford remembered, Kelvin awoke and beamed at him, it can only be because Kelvin did not grasp the significance of Rutherford's discoveries. Atoms, and therefore the earth and sun, contain an internal source of heat. The past temperatures of the earth and sun then are unknowable: either body might have heated up, cooled down, or remained at the same temperature. Rutherford put it this way:

> The discovery of the radio-active elements, which in their disintegration liberate enormous amounts of energy, thus increases the possible limit of the duration of life on this planet, and allows the time claimed by the geologist and biologist for the process of evolution.

According to Rutherford's biographer, Lord Kelvin agreed to a bet that he would soon accept the new radioactivity. At the subsequent meeting of the British Association in 1904, Kelvin paid off. He may have honored the bet, but Kelvin's published statements took the opposite tack. In a series of letters to *The Times* (London) in 1906, asserting that he had spent "more hours in reading the first and second editions of Rutherford's Radio-activity" than almost any other person, Kelvin rejected the whole notion that radioactivity might account for the earth's heat. As for the heat of the sun, that was due not to radioactivity but, as he had long said, to the release of gravitational energy. Kelvin claimed that radium is not a chemical element, but a molecular compound composed of lead and five helium atoms. The heat emitted by radium is not due to alpha particles, he said, but to electrons, which somehow become loaded with energy from the surrounding ether. Kelvin said this process can "go on forever, without violating the law of conservation of energy, and without any monstrous or infinite store of potential energy in the loaded Radium."

Lord Kelvin died in 1907, as far as we can know unrepentant, shortly after attending yet another professional meeting, at which Rutherford found him "as lively as ever." Kelvin leaves historians of science the task of evaluating his paradoxical record. He was right about the all-important second law and a great deal more. But the age of the earth became his demon. By refusing to consider even the possibility that radioactivity called his assumptions into question, Kelvin in essence withdrew from the grand game of science.

The Ideal Hourglass

The thought of using radioactive decay to measure ages occurred to Rutherford almost as soon as he discovered the principles of radioactive decay. He measured the amount of uranium and the amount of helium, which he believed to be the daughter product of uranium, in the sample of pitchblende that he had displayed to his colleague on the McGill campus. Rutherford did not know the half-life of uranium, so he estimated it by using the rate at which the pitchblende emitted alpha particles. Since he knew or could estimate the amount of the parent, the amount of the daughter, and the rate at which parent decayed to daughter, Rutherford had the elements of his hourglass. It was a simple matter for him to calculate the age of the specimen: 700 million years. He went on to measure the age of other minerals and also obtained ages in the hundreds of millions of years. Not surprisingly, given all that he did not know and had to guess or estimate, his ages were inaccurate. But they were far older than Kelvin's and allowed, as Rutherford said, the biologist and geologist the time they needed.

One source of error in Rutherford's uranium-helium method is that the unbonded, gaseous helium atoms often leak out, leaving fewer than should be present and lowering the calculated age. Rutherford knew that an American chemist, Bertram Boltwood (1876–1927), had deduced that lead is likely to be the end product of the decay of uranium and radium. Considering the possible errors in the uranium-helium method, Rutherford suggested that lead might be better for age measurements than helium, for lead has "no possibility of escape" from crystal lattices.

To Rutherford, as to Kelvin, the age of the earth and its constituent rocks and minerals were of secondary interest. Kelvin was intent on correcting the great mistake of British popular geology—belief in Lyellian uniformitarianism—and the age of the earth and sun provided the way. Rutherford wanted to use radioactivity to explore the atom. Once he and his colleagues had described the basic principles of radioactivity and provided a few examples of its utility, Rutherford moved on.

Rutherford had ahead of him the 1908 Nobel Prize—not in his chosen field of physics, but in chemistry. His greatest achievement was the Rutherford model of the atom, with its central nucleus of protons and neutrons surrounded by shells of electrons. Niels Bohr (1889–1962), his distinguished Danish collaborator, said that "Rutherford's achievements are so great that they provide the background of almost every word that is spoken at a gathering of physicists." In 1914, he was knighted as Baron Rutherford of Nelson, his old home town on the South Island.

One of the first to pick up where Rutherford left off was R. J. Strutt, professor of physics at Imperial College in London. Strutt measured the amounts of uranium and helium in uranium-rich bones. Though he found ages as great as 140 million years, they did not correlate with the relative geological ages of the specimens. Apparently, the bones had leaked helium. Strutt began to search for a uranium mineral that would retain helium, and soon settled on zircon, a refractory mineral that is still the primary one used for age dating based on the decay of uranium to lead. An early Paleozoic zircon gave Strutt a uranium-helium age of 321 million years.

Mineral ages interested the American chemist Boltwood only as they helped to show whether, as he and Rutherford suspected, lead is the end product of uranium decay. Boltwood found support for this thesis in a comparison of the amount of uranium with the amount of lead in forty-three mineral specimens: for a given amount of uranium, the older the mineral, the more lead it contained. Using an estimate for the decay rate of uranium, Boltwood calculated ages ranging from 410 million years to the then amazing 2.2 billion. He indirectly became the first to use uranium and lead, rather than uranium and helium, to measure rock ages.

Even the first, inaccurate attempts to calculate geologic ages using radioactivity gave results well beyond Kelvin's limit. Biologists and geologists now had plenty of time; indeed, some of them thought, the new physicists were providing them more time than they really wanted.

Too Much Time?

With Kelvin's assumptions and conclusions invalidated, with mineral ages coming in at hundreds of millions of years, one would have thought that his opponents would have been satisfied. But no. Some ignored the now expanded time scale; others reacted as negatively as they had to Kelvin's scale. Having fought to raise the limit from 24 million years back closer to the 100 million or so that their hourglasses had given, many geologists saw no need to go farther. They understood that radioactivity might invalidate not only Kelvin's methods, but their own.

Using his new method of sodium accumulation, which had received only acclaim, John Joly had determined the age of the earth to be 89 million years and had courageously defended the figure against Kelvin's lower one. Joly had begun as a physicist and then moved into geology, so he understood the findings of the physicists. In 1909, in an attempt to salvage the salt clock, he raised a set of valid questions that well summarize the assumptions inherent in radioactive age measurements.

1. Can one assume that at the time it formed, a mineral contained no daughter product?
2. Can one assume that in the millions of years that have passed since it formed, a mineral neither gained nor lost parent or daughter?

And finally—the point to which Joly was to return until the end of his life:

3. Can one assume that the rate of decay has been constant?

We can translate these into the three conditions necessary for the parent-daughter methods to work: (1) Absence of original daughter; (2) maintenance of a closed system; and (3) constancy of decay rate.

The geological clocks based on stratigraphy and salt, which gave ages consistent with Kelvin's original 100 million years, and the radioactive methods, which gave much older ages, could not both be right. If the radioactive ages were correct, the geological ones would have to be off by a factor of 5, or 10, or 20, and Joly saw no reason why they should be. To accept the ages from radioactivity would have meant admitting that the decades of work that geologists had poured into inventing and refining their clocks had been for naught. Naturally they concluded that, just as Kelvin had been wrong, so might the new crop of physicists be wrong.

Joly noted that while the stratigraphic and salt clocks gave the same age for the earth—approximately 100 million years—the uranium-helium and uranium-lead methods gave different ages for the same minerals. A simple explanation of the discrepancy is that decay rates have varied—the third problem identified above. If decay rates can vary, laboratory experiments might show it.

Madame Curie, Rutherford, and others had tried to discover the source of radioactivity by seeing if they could alter its rate artificially. Curie found that, regardless of the chemical compound or mineral in which uranium was located, it decayed at the same rate. In 1907, Rutherford heated up some radium "emanation" (later found to be radon) to a temperature of 2,500 degrees Celsius and a pressure of over 1,000 bars. The radium emanation ticked away as before. A few years later, Madame Curie conducted the opposite experiment: she froze a radioactive material to the temperature of liquid hydrogen. Others tried magnetic fields and artificial gravity; but the rate of decay never varied. Radioactivity thus appeared to be not a chemical phenomenon, but one that derived from deep within the atom.

If chemistry and physics offered no support for Joly's claim that the rate of decay had varied, he thought geology might. For years, scientists had noticed that microscopic thin sections of minerals, such as mica, often show rings of discoloration, concentric around a tiny inclusion, commonly the mineral zircon. The discovery of radioactivity explained these haloes. Each radioactive element gives off alphas of a certain energy. They go shooting off through the crystal lattice of the enclosing mineral until they slow down and are absorbed; their energy of motion converts to heat. At a distance that corresponds to the energy of the particular alpha, a dark absorption halo forms. Joly correctly reasoned that the darker the halo, the longer the alphas had been absorbed and the older the specimen.

Joly produced artificial haloes and compared them with natural ones. He found that the number of alphas required to produce the observed amount of darkening was far greater than would have been generated over the geological age of the specimen estimated from his short time scale. Two explanations were possible: (1) the geological age was correct, in which case the rate of radioactive decay must have been higher in the past; or (2), the age from radioactivity was correct, in which case the geological ages were far too low. With no independent evidence to lead him to abandon years of work, Joly chose to accept the geological ages and to conclude that the rate of decay had varied. Until his death in 1930, he continued to try to show that the radioactive methods were in error. Like Lord Kelvin, Joly died without ever accepting publicly that the earth is far older than his salt clock had shown him.

Emerging from the Mists

While Rutherford was making his seminal discoveries, a student of Strutt's, a young Englishman named Arthur Holmes (1890–1965), switched his studies from physics to geology. In his first paper, written in 1911 at the age of twenty-one, Holmes reevaluated Boltwood's uranium-lead ages and added several new analyses. It was already apparent to him that

> Wherever the geological evidence is clear, it is in agreement with that derived from lead as an index of age. Where it is

obscure . . . the evidence does not, at least, contradict the ages put forward.

In 1913, Holmes published *The Age of the Earth,* in which he carefully reviewed the debates of the preceding fifty years. Holmes had learned enough geology to be able to assess each of the geological methods and to describe their underlying assumptions and contradictions. He dispatched Kelvin and set the stage:

With these discoveries the long controversy was finally buried, and Kelvin's treatment of the problem was proved to have been fallacious. The discovery of radium did not only destroy the validity of the older thermal arguments; but also, it led directly to the elaboration of a new and more refined method . . . every radioactive mineral can be regarded as a chronometer registering its own age with exquisite accuracy. Indeed, if our interpretation is correct, some of the oldest Archean rocks must date back 1600 million years.

Holmes then summed up the dilemma that had confounded Joly:

From the mists of controversy which for half a century have hung over the subject . . . two methods alone emerge. One of them must be rejected. Which is it to be? Uniformity . . . is involved equally in both calculations. If we favor the uniformity of geological processes—a well-worn doctrine which has done good service—then we must reject uniformity of radioactive disintegration.

If the present is the key to the past, as uniformitarianism would have it, present rates of geological processes extend back indefinitely and the geological hourglasses must be correct. But then the rates of radioactive decay cannot be constant. As Joly had concluded, science would have to make a choice; only one kind of uniformity could prevail.

After reviewing the claims of Joly and others, Holmes came to his conclusion: no independent evidence exists to show that the rate of radioactive decay has varied. Therefore, "The discordance between the time-estimates drawn from the rates of geological and radioactive changes cannot be held to constitute a sufficient reason for rejecting current opinions unless it is conclusively demonstrated that the geological estimates are beyond question." Since "the modern hour-glass is running at two-and-a-half to four times its average rate," Holmes concluded that the geological calculations were off by the same factor.

Soon after World War I, the British Association for the Advancement of Science and the American Philosophical Society held conferences to discuss the age of the earth. By this time, many ages in the hundreds of millions of years, and older, had been measured. Both conferences concluded that the earth's age is about 1.3–1.5 billion years. In the second edition of his book, published in 1927, Holmes said goodbye to the geological hourglass methods. They were, he said, "incapable of providing exact results because the assumption of uniform rates throughout the past cannot be granted."

In 1931, the U.S. National Research Council appointed a committee of distinguished American geologists from a variety of disciplines, augmented by Arthur Holmes, to review the age of the earth. Holmes devised a set of rigorous standards that each dated sample would have to meet to be included in his final tabulation: a mineral would have to be unaltered, of known stratigraphic age, and have been dated by the uranium-lead method. Only a handful of mineral samples passed Holmes's fine screen. From this small but high-quality set, Holmes concluded for the committee:

> No more definite statement can therefore be made at present than that the age of the earth exceeds 1460 million years, is

> probably not less than 1600 million years, and is probably much less than 3000 million years.

By the 1930s, Kelvin was long departed and Joly, the last holdout for the geological methods, was also gone. A new generation of scientists, with no allegiance to their Victorian predecessors, had plumbed the atom and placed their hands on the controls of new and more powerful instruments. The stage was set to provide a precise answer to the question that had puzzled scholars all the way back to Telliamed: How old is the earth?

3

The End of the Debate

The time has gone by when the physicist prescribed dictatorially what theories the geologist might be permitted to consider.

—A. S. Eddington

Unlocking the Atom

Although Rutherford left behind his early interest in measuring mineral ages, his subsequent research, along with that of the other pioneers, led to a series of surprising discoveries that turned out to be crucial to establishing the true age of the earth. These fertile early years were an era of little science—research conducted on a benchtop at minor expense with apparatus that today seems not much advanced beyond string and sealing wax. But with only string and sealing wax, Rutherford was a virtuoso.

Rutherford demonstrated that the atom is mostly empty space but has a positively charged, heavy nucleus that repels the like-charged alphas. With this insight, he invented a model of the atom in which negatively charged electrons orbit the nucleus, like planets around a sun. Niels Bohr extended Rutherford's model to show that electrons travel in orbitals at discrete distances from the nucleus. When an electron moves from one orbital to another, it releases energy with a wavelength characteristic of the chemical element.

In 1920, Rutherford found that when alphas strike a nitrogen atom, their interaction engenders a subatomic particle identical to a hydrogen atom. He concluded that the particle is one of the elementary set that comprise the atom, and therefore all matter. Rutherford named the new particle the "proton."

Working in Rutherford's laboratory at Cambridge, James Chadwick discovered a new elementary particle with nearly the same mass as the proton but with no electrical charge. He christened it the "neutron." For this discovery, Chadwick received a knighthood and the Nobel Prize in Physics.

Rutherford's colleague, Frederick Soddy, found that, whereas the lead in uranium minerals has an atomic weight of just over 206, the lead in thorium minerals is heavier, weighing close to 208. Soddy concluded that lead, and other elements, exists in varieties with the same chemical properties but different atomic weights. He called these different varieties "isotopes," from the Greek for "equal place." The lead in a given mineral is a mixture of different isotopes. Lead from ores rich in uranium has more of the lighter isotopes; lead from thorium ores has more of the heavier isotopes. Prior to the invention of the mass spectrometer, with no way to separate and study the individual lead isotopes, this was as far as the physicists could go.

Transmutation

In the early days of the twentieth century, as revealed by such names as "radium emanation" and "thorium X," Rutherford and his colleagues had detected a baffling set of decay products from uranium and thorium. These fertile daughters made grandparents of their parents by decaying to some other substance. The grandchildren then repeated the process by decaying to great-grandchildren, and so on through many generations. Here was nature at its most mystifying.

By dogged persistence and the ingenious use of their string and sealing wax, the pioneers solved the mystery. Radium emanation and each of the other strange offspring are isotopes that typically are both the product of radioactive decay and are themselves radioactive. Uranium does not decay directly to lead; rather, for example, ^{238}U decays to ^{234}Th (thorium), which decays to another isotope, which in turn decays to another, and so on through a chain of fourteen decay reactions until at last the decay reaches ^{206}lead. Lead-206 is stable, and there the process stops. Radium and radon are among the several intermediate products in the chain. ^{235}U and ^{232}Th have similar decay schemes.

At first it would appear that these complex decay chains, with their many intermediate products, some of them gases that could easily escape, would invalidate the lead methods of age determination. But, in part because the half-life of each of the intermediate reactions is so short, the process works as though uranium decayed directly to lead.*

By the mid-1930s, scientists had deduced the essential facts of the decay of uranium and thorium to lead: ^{235}U decays to ^{207}Pb with a half-life of 704 million years; ^{238}U decays to ^{206}Pb with a half-life of 4,470 million years; ^{232}Th decays to ^{208}Pb with a half-life of 14,000 million years. Lead has one additional naturally occurring isotope: ^{204}Pb. It is neither radioactive nor radiogenic (produced by radioactive decay).

Since some minerals contain neither uranium nor thorium, scientists needed to find methods based on the decay of other parent atoms. To be useful, a parent isotope must be present in a variety of rocks at detectable levels. So that all of the parent will not have decayed away early in earth history, its half-life must be in the hundreds of millions

---

*Books on radiometric dating such as Dalrymple (1991) and Faure (1986) explain this point.

or billions of years. One that meets the requirement is potassium 40, which decays to argon 40 with a half-life of 1.3 billion years. Another is rubidium 87, which turns into strontium 87 with a half-life of 49 billion years. These four parent isotopes—two from uranium and one each from rubidium and potassium—more than suffice to date the earth and the solar system. (The thorium clock is seldom used.)

Parent-Daughter Clocks

Joly, Holmes, and their contemporaries recognized that accurate dating required three conditions: absence of original daughter; maintenance of a closed system; and constancy of radioactive decay rates. They knew that the first two conditions are often violated. For example, the minerals used in uranium-lead dating usually contain ^{204}Pb. Since ^{204}Pb is non-radiogenic, it must have been present when the mineral formed. But the isotopes of an element do not occur alone; therefore, finding ^{204}Pb means that all the other lead isotopes must have been present as well. As to the closed system requirement, early attempts to use the uranium-helium method had shown that it was often honored in the breach. Later work established that under the right circumstances, each of the daughter atoms used in radiometric dating can escape. Instead of accepting these potentially fatal flaws and abandoning attempts to date rocks, pioneering scientists invented clever ways to get around both.

But what of the third requirement: that the rate of radioactive decay remain constant? Madame Curie, Rutherford, and others had been unable to alter decay rates, and neither has anyone since. Because radioactivity is not a chemical process, but stems from within the nucleus, decay rates do not change. Compared to the atom as a whole, nuclei are tiny and protected from interaction by their palace guards of surrounding electrons.

Though no experimental or theoretical evidence indicates that decay rates change, it is possible to test the claim: Measure the age of a rock or meteorite by several different radiometric methods and compare the results. If each method, based as it is on a different parent-daughter pair, gives the same age, then decay rates have not varied.

The Beauty of a Straight Line

The rubidium-strontium decay system is simpler than uranium-lead, and the principles are the same. Rubidium is an uncommon element that does not form its own minerals, but substitutes for potassium, which is common. Every so often, in the nucleus of an atom of ^{87}Rb, a neutron converts to a proton and in the process the nucleus ejects an electron. The loss of a neutron, if uncompensated, would lower the mass of the atom by 1, but the gain of a proton would raise it by 1, so the mass stays the same at 87. The gain of a proton raises the atomic number of the decaying atom by 1. Since atomic number determines the nature of a chemical element, the decay converts the atom of rubidium into an atom of the next element above it in the periodic table, strontium.

Strontium has four naturally occurring isotopes. The other three besides ^{87}Sr, like ^{204}Pb, are not produced by radioactive decay (are non-radiogenic). Neither are they radioactive. When analyzing strontium with a mass spectrometer, all four isotopes are routinely scanned and their abundances measured. One of the non-radiogenic isotopes, ^{86}Sr, typically has about the same abundance as ^{87}Sr, so it is convenient to use the ratio of ^{87}Sr/^{86}Sr to express the degree to which the strontium is radiogenic. The higher the ratio of ^{87}Sr/^{86}Sr, the more of the strontium that came from radioactive decay, and conversely.

The other piece of information needed to calculate an age is the amount of the parent: ^{87}Rb. After all, a substance with a large amount of radiogenic strontium—with a high ratio of ^{87}Sr/^{86}Sr—could be young or old; it depends on whether the mineral has a little or a lot of the parent. The less rubidium a mineral contains, the older it must be to have generated a given amount of radiogenic strontium, as expressed by the $^{87}Sr/^{86}Sr$ ratio.

We can describe the decay of ^{87}Rb to ^{87}Sr in a mathematical equation. We can divide both sides of the equation by the same factor without changing the relationships. If, on the daughter side of the equation, the amount of ^{87}Sr is divided by the amount of ^{86}Sr to give the ratio of $^{87}Sr/^{86}Sr$, then, on the parent side, the amount of ^{87}Rb must be divided by the amount of ^{86}Sr, to give the ratio of $^{87}Rb/^{86}Sr$. This is easiest to see on a graph (Figure 3.1).

After the mass spectrum of the strontium and the amount of rubidium in a sample are measured, the two ratios, $^{87}Sr/^{86}Sr$ and $^{87}Rb/^{86}Sr$, are calculated and plotted, as shown in the figure below. (Think of this graph as a plot of the relative amount of the parent on one axis against the relative amount of the daughter on the other.) In one stroke, the

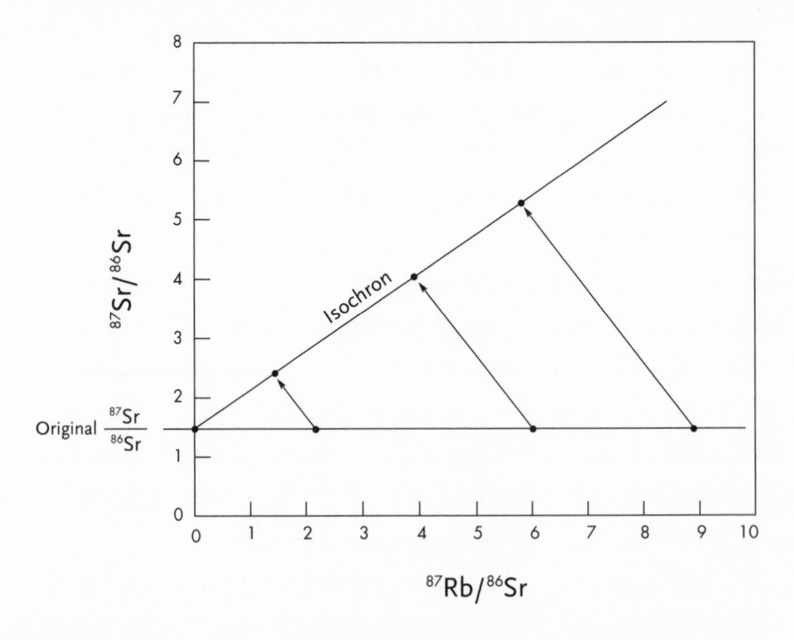

Figure 3.1 Ideal Rubidium-Strontium Isochron. Data points start on the horizontal line of original $^{87}Sr/^{86}Sr$ and over time move up and to the left. If neither parent nor daughter atoms are gained or lost, the points always lie on a straight line. The slope of that line reveals the age of the material, the steeper the older.

graph satisfies two of the three conditions for radiometric dating: the absence of original daughter and maintenance of a closed system.

Imagine a magma cooling beneath the earth's surface. Because the strontium isotopes are identical chemically, each part of the magma will have the same ratio of $^{87}Sr/^{86}Sr$. As different minerals freeze out, each will have the same $^{87}Sr/^{86}Sr$ ratio as the overall magma. Since some minerals take up more rubidium than others, each mineral will have a different amount of the parent, rubidium, relative to the amount of strontium. In other words, each will have a different ratio of $^{87}Rb/^{86}Sr$. Therefore, each mineral will subsequently generate ^{87}Sr at a different rate. This mathematical reasoning leads to the following conclusion: If a set of minerals starts out at the same time with the same ratio of $^{87}Sr/^{86}Sr$, but with different ratios of $^{87}Rb/^{86}Sr$, and if the set remains a closed system, at any later time the points will fall on a straight line on the rubidium-strontium graph. Since the line defines the locus of points that have the same age, we call it an isochron.

The isochron is the familiar straight line from geometry: $y = mx + b$, where m is the slope and b is the intercept on the y ($^{87}Sr/^{86}Sr$) axis. The intercept gives the original $^{87}Sr/^{86}Sr$ ratio. The age calculation does not require the original ratio because the mathematics of rubidium-strontium decay show that the age to which the isochron corresponds depends only on its slope. The steeper the slope, the greater the age. Calculating a precise age requires knowing only the slope of the isochron; whether the substance had original strontium, and if so how much, does not matter. Not only the rubidium-strontium method, but several others, use the isochron. One of the fundamental conditions of radiometric dating—the absence of original daughter—the isochron renders irrelevant. One down, two to go.

Next, consider how the isochron reveals whether the system has remained closed. Except by coincidence, a group of data points can fall on a single isochron only as a result of the set of circumstances just described. Each analyzed sample had to have formed at the same time, with the same original $^{87}Sr/^{86}Sr$ ratio, and have neither gained nor lost parent or daughter. The data point for a sample with a different age or initial ratio of $^{87}Sr/^{86}Sr$, or that had gained or lost parent or daughter, would fall on the isochron only by chance.

Thus, if the data points on an isochron diagram closely fit a straight line, the samples have remained closed. If the points scatter, either the system has opened or the samples do not belong together. In that case, the "age" is meaningless. Two down, one to go.

All this may be rather dry, admittedly. To measure the abundances of the rubidium and strontium isotopes using a mass spectrometer is tedious. But for a scientist, the isochron diagram is closer to a miracle than a mere scientific device. To witness the hard-won data points begin to align themselves, one after another, along an isochron, is like witnessing the emergence of a beautiful butterfly from a drab cocoon. The linearity of an isochron on an ancient material reveals a stark cause and effect to which, through the eons, nothing has mattered but the steady, unalterable transformation of parent into daughter. Continents have danced across the surface of the earth, mountains have emerged and been eroded to dust, oceans have opened and closed, species have appeared and gone extinct, civilizations have arisen and vanished. No matter: the inexorable hand of the clock of radioactivity ticks, and, having ticked, moves on.

Paradoxically, time is incomprehensibly long, yet measurable. Geochronologists are the Lilliputians, time the Gulliver whom they ensnare and vanquish with their invisible strings: the isotopes.

Pioneers

The rubidium-strontium method came of age only in the 1950s. Prior to that, efforts to measure the age of the earth relied on the more complicated but more informative lead isotopes. In the early 1930s, Alfred Nier of the University of Minnesota built the first mass spectrometer able to separate and measure the lead isotopes. He found that the abundances varied considerably in nature, but that the more ^{206}Pb a mineral had, the more ^{207}Pb it had. Since both are daughters of uranium, Nier deduced that lead in nature is a mixture of the primordial lead with which the solar system and the earth began, together with radiogenic lead produced by the decay of uranium and thorium since. Therefore, the lead in each rock and mineral combines primordial and radiogenic lead in some proportion.

Among the radioactive parent-daughter pairs, lead is unique in having three clocks running at once: two from uranium decay and one from thorium decay. From the mathematics of lead decay, it turns out that if one knows the abundance of three of the four lead isotopes (204, 206, and 207) in some material today, and if one knows their abundance in primordial lead, one can calculate the age of the material from that information alone. One does not even need to know how much parent was present.

In 1941, Nier and his colleagues published analyses of lead in minerals rich in uranium and thorium, as well as in several galenas, a lead sulfide mineral that contains none of either parent. Making reasonable assumptions, Nier calculated ages for some of the uranium and thorium minerals of over 2 billion years, but did not attempt to derive the age of the earth. Though a world war intervened, three scientists, each in a different country, found a way to use Nier's data to calculate at least a minimum age for the earth.

Nier's paper appeared in the July 15, 1941, issue of the Western journal, *Physical Review.* By late 1941, advance units of the German Army had reached the outskirts of Moscow and could see the domes of the Kremlin less than 25 miles away. On October 20, 1941, Russian authorities declared the city in a state of siege. Nevertheless, on February 20, 1942, the Doklady of the Russian Academy of Sciences received a manuscript from Academician E. K. Gerling of the Radium Institute of Moscow entitled, "Age of the Earth According to Radioactivity Data."

One of Nier's galenas, from Ivigtut, Greenland, had the least radiogenic lead yet measured, placing it as close as one could get at the time to the primordial lead with which the solar system and the earth began. Gerling calculated that the time required for lead to evolve from the Ivigtut composition to the more radiogenic lead of one of Nier's youngest galenas would have been 3,230 million years. The same calculation on another galena gave 3,950 million years. Gerling knew that since the Ivigtut galena is not as old as the earth, its lead is not primordial, and therefore he knew that his ages were only minima. He concluded that the age of the earth "is not under 3,000–4,000 million years."

One can only wonder how, during the height of the German-Russian War, his city under siege, Gerling could have gotten hold of Nier's paper in the first place. Once he had, how could he so quickly have understood how to use Nier's data? How did he find the concentration to put the data to work? However he did it, Gerling raised the standard: a respectable scientific journal could now report an age of 4 billion years.

A now more senior Arthur Holmes had not lost interest in the age of the earth. With few students to occupy him in his post as professor at the University of Durham, Holmes spent the war years in research and writing. During 1946–48, he published a series of papers in which he employed Nier's data and a variation of Gerling's method (which he did not know about) to estimate the age of the earth. Using a complex approach that few could have understood, Holmes concluded that "the most probable age of the earth is about 3,350 million years."

The third in the trio of great scientists who, working independently during and just after the worst war in history, knew what to do with Nier's data was a German physicist named Friedrich G. Houtermans (1903–1966). One of the most remarkable characters in the history of science, Houtermans could have stepped from the pages of a spy thriller. Before the war, in sympathy with the expressed ideals of the Soviet Union, Houtermans immigrated to Kharkov. This allowed first Stalin's NKVD, and after the Germans invaded, the Gestapo, to arrest and torture him.

Like Holmes, Houtermans was usually ahead of his time. In the early 1930s, he had pointed out the possibility of a self-sustaining nuclear chain reaction, the basis for the later atomic bomb. In August 1941, he published a report called "On the Question of Unleashing Chain Nuclear Reactions." In it, Houtermans identified plutonium as a more likely substance than uranium for chain reactions, the same conclusion that Manhattan Project scientists later reached. In December 1942, the Chicago team of nuclear researchers, led by Enrico Fermi, had yet to produce a controlled chain reaction, when they received a cable from Switzerland. It said only: "Hurry up. We are on the track." The wire had come from Houtermans through an intermediary. The Germans never did get the bomb.

Immediately after the war, Houtermans wrote a series of papers on the age of the earth, the first of which appeared in 1946 and was entitled "The Isotopic Abundances in Natural Lead and the Age of Uranium." Houtermans plotted Nier's data on a new kind of diagram in which one axis represented the ratio of $^{207}Pb/^{204}Pb$ and the other the ratio of $^{206}Pb/^{204}Pb$. As with rubidium-strontium, leads that start out at the same time with the same primordial abundance, and that remain closed systems, will at any later time plot along a straight line on this diagram. Houtermans called the line an isochron, the first use of the term. The key facts are the same as with the rubidium-strontium isochron: the age of a set of specimens depends only on the slope of their Houtermans lead isochron; how well the points fit the line reveals how well the closed system requirement has been met.

Plotting Nier's data, Houtermans obtained an isochron whose slope corresponded to an age of 2,900 million years. He thought this was either the age of uranium—in other words, the age of the elements themselves—or the age of the earth's crust.

The calculations of Gerling, Holmes, and Houtermans could give no more than minimum ages because the Ivigtut galena is not primordial, and, for all they knew, might be a long way from it. Until the composition of primordial lead could be determined, the true age of the earth would always remain out of reach. But where on the earth, with its complex history, to find lead that still retains its original isotopic abundances? In 1947, Houtermans made the prescient suggestion that one need not search the earth for primordial lead—meteorites that fall from the sky are likely to preserve it.

Iron meteorites contain minerals that, like galena, have no detectable uranium. If these meteorites have been closed systems, isolated in space since the beginning of geologic time, they may preserve the primordial lead of the solar system. Harrison Brown of the University of Chicago, mentor to the next person in the story, independently made the same suggestion as Houtermans in the same year.

In different countries, minds converged to the same conclusion: To measure the age of the earth, use the primordial lead in iron meteorites. But in the aftermath of a devastating world war, only the United States had the scientists, the methods, the instruments, and the funds to do the experiments.

Duck Soup

Both Houtermans and Brown understood that, given the widespread occurrence of uranium as a trace element, it is highly unlikely that any terrestrial rock or mineral preserves truly primordial lead; instead, all have acquired at least some radiogenic component. Brown had worked on the Manhattan Project, one of whose major tasks was to separate ^{235}U from ^{238}U. In order to determine how well they had separated the two isotopes, scientists analyzed their abundances with a mass spectrometer. When Brown returned to the University of Chicago, he knew that a "mass spec" could also measure the isotopic composition of lead in meteorites.

Brown, inept to the point of endangerment in the laboratory, was not the person to carry out the analyses. He excelled instead at identifying an important research problem, conceiving a solution to it, and then finding both the funding and someone else, usually a graduate student, to do the work. To make the meteoritic lead measurements, Brown needed a meticulous, hardworking, never-say-die young scientist. He found exactly such a person, a graduate student named Claire Cameron Patterson, and assured him that the problem would be "duck soup." Brown predicted that the research would make Patterson famous as the person who had finally measured the age of the earth.

If the life of Friedrich G. Houtermans reveals the nightmare of being a scientist in not just one but two of the worst totalitarian dictatorships in history, the life of Claire Patterson exemplifies the best years in the life of a typical American scientist who came of age just before World War II. Patterson (1922–1995) grew up on a farm in Iowa and attended one of the nation's fine liberal arts colleges, Grinnell, in his home state. One of his classmates was Doris Nininger, whose father was the most successful collector of meteorites. In partial payment of Doris's tuition, Nininger gave some of his meteorite specimens to Grinnell College, where they resided on a shelf in Patterson's classroom. One was a fragment of an iron meteorite from Meteor Crater, Arizona. Meteorites receive the names of the site at which they are collected; those from Meteor Crater take theirs from the nearby trading post at Canyon Diablo. After the war, Patterson was to encounter Canyon Diablo again and it was to make him famous. Duck soup the project was not.

After graduating from Grinnell with a degree in chemistry, having first suffered a two-week expulsion for sins against the college, in 1944, Patterson went to the University of Chicago to work on the Manhattan Project. From there, he and his young wife, also a Grinnellian chemist, transferred to the Oak Ridge National Laboratory in Tennessee, where he learned to use a mass spectrometer. Like many others, Patterson agonized for the rest of his life over his role in producing the atomic bomb.

After the war, the Pattersons returned to the University of Chicago to join a distinguished cast of luminaries from the Manhattan Project. They included Harold Urey (1893–1981), the chemist who won the Nobel Prize in 1934 for discovering deuterium, the isotope of hydrogen with one proton and one neutron and thus twice the mass of regular hydrogen. Willard Libby was there—he was to win the prize in 1960 for his invention of carbon-14 dating, based on work done right after the war at Chicago. So was Patterson's mentor, Harrison Brown.

The research that Brown predicted would be duck soup took seven years and required the building of an entire new laboratory and the invention of novel techniques. Brown had based his optimism as to how long the work would take on chemical analyses that showed that iron meteorites contain enough lead to make it easy to separate and analyze. But Patterson soon discovered that industrial lead is so ubiquitous in the environment, and the effects of its contamination so difficult to remove, that almost all previously reported lead abundances were erroneously high. Instead of measuring the lead inherent in a material, these earlier analyses had usually recorded the much higher level of contaminant lead. The discovery of the extent of lead contamination in the environment eventually changed Patterson's career and our health.

Patterson and Brown moved their research to the California Institute of Technology and the then even more lead-contaminated environment of Southern California. In order to measure lead and uranium in meteorites, Patterson had to reduce the effects of environmental lead contamination to the absolute minimum. At Cal Tech, he invented the "clean lab," which lowered contamination and allowed him to measure extremely low concentrations of lead and uranium.

By 1953, Patterson and his colleagues reported that the isotopic composition of lead in one of the minerals in Canyon Diablo, a uranium-free iron sulfide called troilite, was the most primordial yet measured. Houtermans, in postwar Germany, leaped on Patterson's data. Before the year was out, Houtermans published an article in which he concluded that, if one assumes that the lead in selected young lead ores started out not with the Ivigtut composition, but with the more primitive abundances of Canyon Diablo, the corresponding age of the earth is 4.5 ± 0.3 billion years. Patterson reported the same conclusion at a conference held in September that year. Thus, in 1953, it became clear that if it had begun with the lead in Canyon Diablo, the earth was 4.5 billion years old.

The modern figure for the age of the earth derives from Patterson's classic 1956 paper, "Age of Meteorites and the Earth." Patterson analyzed three stone meteorites, which contain appreciable uranium and therefore have evolved a radiogenic lead component, and two iron meteorites, one of them from Canyon Diablo. Patterson's data points, reproduced in Figure 3.2, fell along a nearly perfect lead isochron. From its slope, Patterson calculated an age of 4.55 ± 0.07 billion years for the group of meteorites. This he equated with the age of the earth.

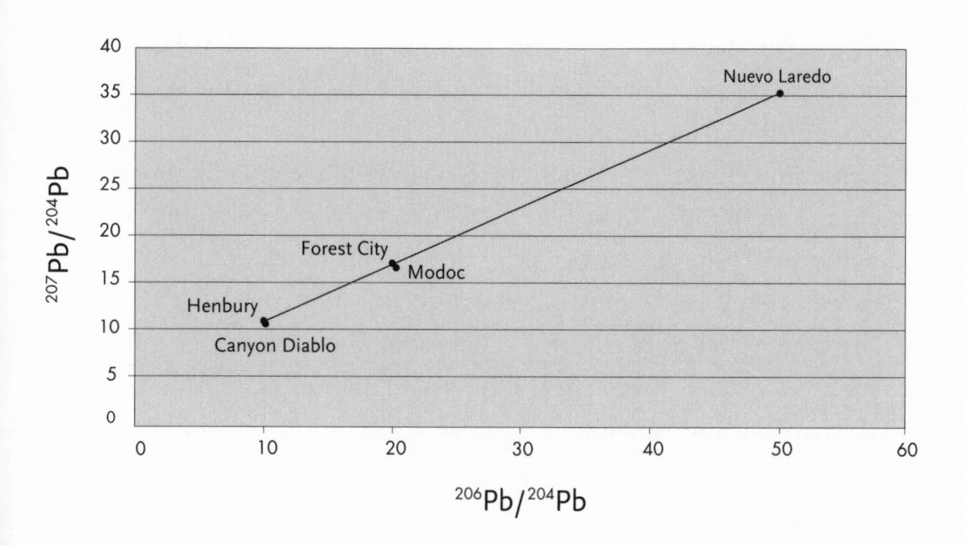

Figure 3.2 Patterson's Meteorite Lead Isochron (after Patterson, 1956). The slope of the line corresponds to an age of 4.5 billion years, which Patterson concluded was the age of the earth and the solar system.

Patterson's life and career serve as a metaphor for twentieth-century science. At first he devoted himself to purely scientific questions, but inevitably his attention turned to the implications of his research for environmental pollution and human health. Rather than admit to this, Patterson preferred to play the role of no-nonsense scientist whose goals, as he put it, were nothing but "science, science, science."

Patterson's transformation began with his need to find the source of the ubiquitous environmental lead. He soon discovered high concentrations of lead in the surface waters of the ocean and in otherwise pristine snow. Lead in snow could have gotten there only from the atmosphere. But except for that introduced by the burning of leaded gasoline, the atmosphere has no lead. Patterson then conducted one of the most arduous and clever experiments in modern science. He collected successively deeper layers of Greenland ice and measured their lead concentrations. He found that the amount of lead in the atmosphere, as recorded in the ancient snow layers, had begun to rise at the time of the industrial revolution. But at the moment in the 1920s when gasoline manufacturers began to add lead, the level of lead in the atmosphere jumped alarmingly. It continued to climb until it reached two hundred times its natural level.

Patterson fought for decades to get his findings into the public arena. In spite of opposition from the leaded gasoline industry, his science was so superb that there was no denying his conclusions. Patterson's research was key to the passage of the Clean Air Act of 1970.

An apocryphal story tells of a cautious mathematician who, upon seeing a black sheep in a Scottish field, would only go so far as to conclude that Scotland contains at least one sheep at least one of whose sides is black. Patterson had shown that one small set of meteorites is 4.5 billion years old. How did this most meticulous of scientists have the confidence to take the opposite tack of the overly prudent mathematician and leap from his meager sample to the age of the entire solar system and everything in it?

4

The Age of Meteorites, the Moon, and the Earth

When you cannot measure, your knowledge is meager and unsatisfactory.
—Lord Kelvin

Aliens

Before the space age, scientists had only two types of material from the solar system to date: meteorites, and the rocks and minerals of the earth. The formidable Patterson had established the age of two iron and three stone meteorites, a small sample indeed. That age he equated to the age of all meteorites, the earth, and the solar system itself. Patterson had the confidence to stake his career on that leap because he and other scientists had convincing evidence that meteorites date back to the beginning of the solar system.

Meteorites must have inspired curiosity since the dawn of intelligence. Who has not seen a shooting star, the tiny meteors that enter the earth's atmosphere and blaze for an instant before burning up? The brighter skies of earlier times brought the heavens close up and must have provided a much more impressive light show than we can witness today.

Since, to our predecessors, the heavens appeared empty, it was reasonable to conclude that meteorites were rocks carried up into the sky, perhaps by waterspouts, from whence they fell to earth again. In 1769, the august French Academy of Sciences appointed the chemist Antoine-Laurent Lavoisier to head up a commission to study the provenance of a stony meteorite collected the year before. The academicians pronounced it an earthly rock.

In the 1790s, a German lawyer and scientist named E. E. F. Chladni described two iron meteorites, one a 13,500-kilogram giant from Argentina, that were impossible to mistake for earthly rocks. Chladni evaluated the different possible origins of meteorites and concluded that they were foreign to the earth and had fallen from the heavens. His careful scientific work had less effect on popular opinion than the hard-to-miss fall of thousands of meteorites on the village of L'Aigle, west of Paris, in 1803. But not everyone was convinced. Upon hearing of the account from two academics of a meteorite fall in Connecticut, Thomas Jefferson allegedly said, "It is easier to believe that Yankee professors would lie, than that stones would fall from the sky." Even if the quotation is apocryphal, Jefferson undoubtedly would have spoken for his contemporaries.

As the only extraterrestrial objects on earth, meteorites have interested so many people for so long that a vast literature exists. Most of what we know about the early history of the solar system comes from the study of meteorites, which are like fossils from the beginning of geologic time. Over twenty thousand meteorites have been discovered, most within the last several decades on the Antarctic ice, where they are easy to spot. Countless more have fallen and been overlooked, or have been eroded to dust.

Stony meteorites are easy to miss because they are composed mainly of the same minerals that make up terrestrial rocks. Few who stubbed their toe on a stony meteorite could have realized that the offending and apparently mundane object had spent more than 99 percent of the history of the solar system soaring through the frigid vacuum of space. Most stone meteorites contain rounded globules called chondrules (after the Greek word for "grain"), which are made of the same silicate minerals that we find in high-temperature rocks believed to come from the earth's mantle.

Iron meteorites do not resemble any earthly rock, but look more like the product of a blast furnace. Due to their odd appearance, meteorite collections are full of irons. In between the stones and irons lie many varieties with a bewildering set of names. One group, the SNC meteorites, come from Mars and thus are the only specimens we have on Earth from another planet.

The abundance of different types of meteorites depends on whether one counts the number that were collected only after they were seen to fall (falls), or the number that were found but not seen to fall (finds). Irons and near irons make up two thirds of finds, but stones comprise 95 percent of falls. Obviously, in space, stones dominate. Canyon Diablo is a classic iron meteorite; three of Patterson's other meteorites are stones that contain enough uranium so that over time their lead has grown more radiogenic.

The astronomer Johannes Kepler (1571–1630) had predicted that a planet might lie between Mars and Jupiter, but instead the region contains only asteroids. Chladni reasonably concluded that the meteorites are the remains of an exploded planet, an idea that lasted to the beginning of the space age. By then it had become clear that the chemistry of meteorites does not fit the exploded planet model. Neither does the mass of the asteroids, which in total amounts only to a small fraction of the mass of our moon. A true planet could not have been so small.

The evidence that meteorites come from within our solar system is overwhelming. Those whose observed trajectories have been tracked backwards were orbiting the sun in the same direction as the planets and in the same plane. They traveled at speeds of 15–20 kilometers per second, the expected velocity for objects originating within the solar system. The farthest point in their orbits lay within or just beyond the asteroid belt. Meteorites are an intimate, fundamental part of our solar system.

Today, the evidence convinces scientists that meteorites form when asteroids of different sizes and histories collide, producing fragments with a wide variety of compositions. Smaller asteroids, up to 200 kilometres in diameter, never became hot enough to differentiate into crust, mantle, and core, but instead remained homogeneous. In contrast, the larger asteroids, up to 600–700 kilometres, had enough internal heat to differentiate. Jupiter's gravity sends some meteorites to Earth, presenting us with a collection of analogs to our own much larger and largely inaccessible planet.

If the asteroids are remnants of the original building blocks of the solar system, and if meteorites come from the asteroid belt, then meteorites have the same age as the solar system, justifying Patterson's giant leap. Scientists began to measure meteorite ages as soon as they invented radiometric dating. They first tried the uranium-helium method, but the resulting ages ranged from 1 million years to 7,000 million, reflecting the mobility of the unbonded helium. By the early 1990s, as summarized in G. Brent Dalrymple's masterful *Age of the Earth,* well over a hundred age measurements had been made on seventy-nine different meteorites. All but ten of the ages exceed 4.0 billion years, with the majority falling between 4.4 and 4.6 billion. Those that give the youngest ages always show independent evidence of having been severely shocked, presumably due to a recent impact on their parent body. Given the consistency of meteorite ages, two meteorites can stand in for them all.

Remember that rocks and stony meteorites are composed of individual components: minerals; and in the case of stony meteorites, the rounded globules called chondrules. Each individual mineral, chondrule, or other component of a rock or meteorite typically has at least a slightly different amount of parent and daughter, as though each were an independent entity. Scientists sometimes separate these individual components, measure the amounts of parent and daughter in each, and plot them up to provide what is called an internal isochron. If the components remained closed systems, the internal isochron gives the same age as would the entire rock. At other times, scientists grind up a piece of the rock or meteorite and plot it along with others to produce a whole-rock isochron. The choice of whether to use minerals or the whole rock depends on how much material is available and whether the question is the age of a single rock or meteorite, or the age of a group of similar meteorites.

In 1969, a rare stone meteorite rich in carbon fell onto the small Mexican town of Allende. Because the chemical composition of the carbonaceous stone meteorites most closely resembles that of the sun, we believe them to be among the most primitive objects in the solar system. The internal lead isochron for twenty-six different mineral and other components of the Allende meteorite is shown in Figure 4.1. The fit is nearly perfect, indicating that each of the components has remained a closed system. The slope of the isochron gives the age: 4.553 billion years, with an error of only ± 0.004 billion years, less than 0.1 percent. Plotted for reference but not used in the calculation is the modern analysis of lead in the Canyon Diablo meteorite. The Allende isochron passes right through the Canyon Diablo point, suggesting that each of the Allende components started with the lead isotope composition of Canyon Diablo and that Canyon Diablo does preserve the primordial lead isotope ratios of the solar system, as Patterson assumed. A second group of researchers, also using the lead isochron method, dated Allende at 4.565 ± 0.004 billion years. Several potassium-argon–age measurements on the meteorite average 4.53 billion years. The nearly perfect linearity of the Allende lead isochron speaks for itself: this meteorite is 4.55 billion years old.

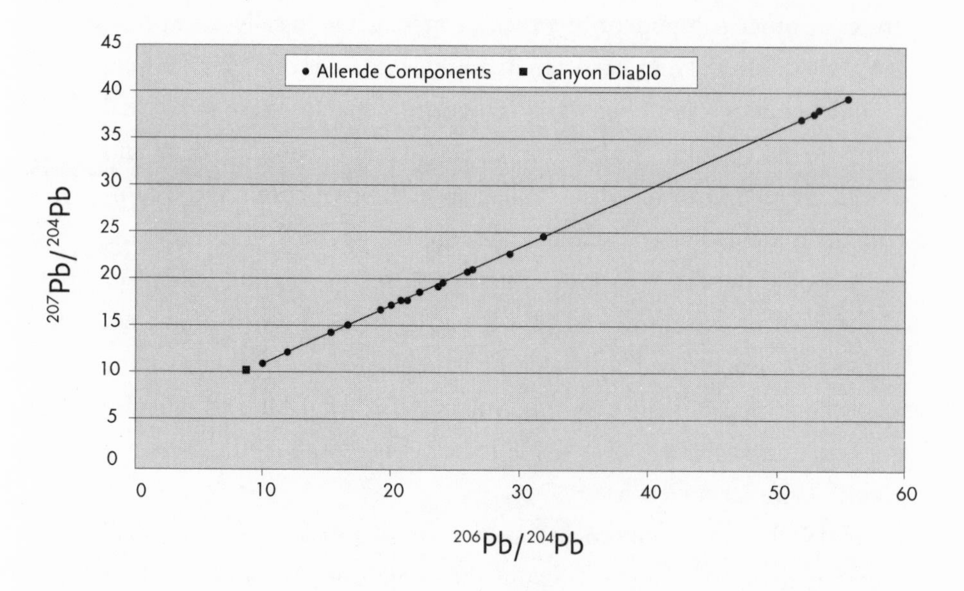

Figure 4.1 Lead Isochron for the Allende Meteorite (after Tatsumoto et al., 1976). The nearly perfect fit of the points to a straight line demonstrates beyond a doubt that Allende is 4.55 billion years old.

Several different methods have been used to date the meteorite known as St. Severin, which comes from western France. Table 4.1 summarizes the results.

The use of the isochron method for any parent-daughter pair satisfies two of the three requirements for accurate radiometric dating: absence of original daughter and maintenance of a closed system. The identity of result for St. Severin using five different parents—one each for rubidium, potassium, and samarium; two for uranium—shows that none of the decay rates changed. If one had changed, that method would have given a different age from the others. It is inconceivable that each would change by just enough to bring the result of five different methods into exact agreement. St. Severin shows that the processes that have been operating in our solar system for 4.5 billion years have not caused the rates of radioactive decay to change. This result—and many other examples exist—validates all the assumptions of radiometric dating. Three down, none to go.

The results for Allende and St. Severin, combined with those from scores of other meteorites, prove beyond doubt that the meteorites formed 4.5 billion years ago. Few scientific facts are better established. Nearly half a century before this book was being written, Claire Patterson, using much more primitive equipment and only a handful of meteorite specimens, obtained an age for meteorites of 4.55 ± 0.07 billion years. In retrospect, he was lucky, for later refinements in half-lives and techniques, any one of which might have invalidated his result, tended to offset each other. In spite of scores of subsequent meteorite analyses, no one has improved on Patterson's age of the earth. The genius of this meticulous and stubborn experimenter needs no greater testimony than the identity of his meteorite age and those of Allende and St. Severin.

Table 4.1 St. Severin Meteorite

Method	Age (billions of years)
rubidium-strontium	4.51
argon-argon*	4.43
	4.38
	4.42
samarium-neodymium†	4.55
lead-lead internal	4.543
lead-lead whole rock	4.549
	4.555
	4.551

*The argon-argon method has replaced the older potassium-argon method. It is less susceptible to loss of argon.

†The samarium-neodymium method is identical in principle to rubidium-strontium.

In 1956, when Patterson published his age of the earth and the solar system, meteorites were the only extraterrestrial specimens available. But within a decade, that would change and Patterson's assertion that the solar system is 4.5 billion years old would undergo a second test: the dating of moon rocks.

Apollo

The dark seas, or maria, of the Janus-faced moon scatter randomly among the brighter highlands. The NASA Apollo missions begun in 1967 found that lunar rocks from both regions had analogs among the rocks of Earth. Basalts from the maria outwardly resemble terrestrial basalts. Rocks from the lunar highlands correspond to a less common earthly rock rich in the mineral plagioclase feldspar.

Well before the Apollo missions began, earth-bound geologists, led by Eugene Shoemaker (1926–1997), an American working at the U.S. geological Survey, applied Steno's principle of superposition and common sense to observations and photographs of the moon. They soon deduced that the mare basalts are younger than the highland rocks, which the dated Apollo samples confirmed. But youth is only relative: with ages ranging from 3.0 to 3.9 billion years, even the youngest lunar basalt returned by one of the Apollo missions is older than all but the oldest terrestrial rocks.

As an example of an ancient lunar basalt, consider specimen 10072, collected by the Apollo 11 astronauts from Mare Tranquillitatis. By the rubidium-strontium method, this basalt presents a nearly perfect isochron corresponding to an age of 3.57 ± 0.05 billion years. By the samarium-neodymium method, basalt 10072 gives exactly the same age, with an even smaller error. The average of six argon dates is 3.55 billion years. Thus, we can say with complete confidence that the age of this lunar basalt is 3.5–3.6 billion years. The identity of the ages from three different methods again validates all the assumptions of radiometric dating.

Basalt is a differentiated rock produced when the interior of a planet melts, which by definition means that its chemical systems were open. To determine the original age of a planet or a moon requires a more primitive, undifferentiated rock, ideally a piece of the original crust that has remained a closed system in spite of all that may have happened to it. On a moon continually bombarded with meteorites, finding such a material may be a tall order; and indeed, only a few lunar samples give ages above 4 billion years, though the sampling is limited, to say the least. But given the validity of radiometric dating, even one lunar specimen as old as the meteorites would make the case. Here consider lunar sample 72417, a rock rich in the mineral olivine (peridot), collected by the Apollo 17 astronauts from the base of the highlands adjoining Mare Serenitatis. The almost perfect rubidium-strontium isochron for this specimen gives an age of 4.47 ± 0.10 billion years. Several other lunar samples also give ages as old. Thus, meteorites and the moon have the same primordial age: 4.5 billion years.

The Oldest Rocks

The earth must be at least as old as its oldest rock. Unfortunately, we cannot proceed by first locating likely candidates for the oldest rock using geological relationships and then dating them using one of the radiometric methods. Not only do Precambrian rocks offer no macrofossils to help establish their relative ages; usually they have been severely folded and metamorphosed. To correlate these ancient rocks based on their stratigraphy—often there is no stratigraphy—or based on rock type, is next to impossible. Prior to the invention of radiometric dating, Precambrian geology, as Winston Churchill said about the Soviet Union, was "a riddle wrapped in a mystery inside an enigma." But, Churchill added optimistically, "perhaps there is a key." For Precambrian geology, the key is radiometric dating.

Typically, geologists make an educated guess based on fieldwork as to which Precambrian terrains are likely to contain the oldest rocks, and then date those rocks using one or more of the radiometric methods. Then they, or their colleagues, search for even older rocks and repeat the process. As a result, by the end of the twentieth century, scientists had dated thousands of Precambrian rocks and found older and older ones. Each continent turns out to have truly ancient rocks, far older than anyone would have predicted.

The ancient rocks of the Canadian Shield have yielded many ages in the range of 2.6 to 2.8 billion years. Until the advent of modern techniques for dating zircons, the oldest North American rocks outcropped south of the border near the small town of Morton, Minnesota. These rocks are so complex that for years it proved difficult to be sure that collected samples actually came from the same rock units. Eventually, geologists worked out the relationships and dated the Morton gneiss (a metamorphosed granite) at 3.48 billion years by the rubidium-strontium method and at 3.59 billion years by uranium-lead.

Since the days of Strutt and the young Holmes, tiny, resistant crystals of zircon have been the most useful minerals for age dating. Zircon's ability to imprison radiogenic lead isotopes without benefit of parole is key. Experience has shown that helium is the first to leave a heated crystal lattice, followed by argon, followed by strontium. But even after these mobile atoms have vacated their crystal premises, the lead in zircon remains. Argon and strontium thus tend to date the time of metamorphic heating, whereas zircon ages often "look back" to the original age of the rock. Not surprisingly, the oldest ages of all have come from zircons.

In the northwest part of the Slave Province of Canada lie some complex rocks called the Acasta gneisses. A group of over fifty zircons from these gneisses, analyzed by the lead method, gave an average age of 3.96 billion years. Further zircon studies published in 1999 found three with ages ranging from 4.0 to 4.03 billion years.

The instrument that produced these dates would have seemed like science fiction to Rutherford at his simply equipped benchtop. The SHRIMP (sensitive high-resolution ion microprobe), developed at the Australian National University in Canberra, allows researchers to measure the isotopic abundances of uranium and lead in individual zircons in only a few minutes. Some zircon grains display what appear to be concentric growth bands; SHRIMP is so sensitive that it can date these microscopic bands individually.

In spite of being difficult to reach and sample, because of their antiquity a suite of rocks from western Greenland is among the most accurately dated. The Isua, Greenland, "supracrustals" give the ages shown in Table 4.2.

Table 4.2 Age of Rocks from Isua, Greenland by Different Methods

Method	Age (billion years)
uranium-lead	3.81 ± 0.02
rubidium-strontium	3.71 ± 0.07
samarium-neodymium	3.75 ± 0.04
lead-lead	3.70 ± 0.07

As with St. Severin and the basalts of Mare Tranquillitatis, not only are the Isua rocks ancient, they give the same age by several different methods, once again validating the assumptions of radiometric dating.

Other continents also have ancient rocks: China, Venezuela, and Zimbabwe each have ones older than 3.5 billion years. A suite of zircons from a metamorphic rock from Enderby Land in Antarctica gives 3.93 billion years. The most intriguing and ancient of all come from the Yilgarn Block of western Australia. One of the rocks in the Yilgarn Block is a quartzite, or metamorphosed sandstone. Sandstones comprise mineral fragments, mainly but not exclusively quartz, eroded from older rocks. They often contain minor zircon. During the 1980s, scientists at the Australian National University found that most of the zircons from the Yilgarn quartzite gave 3.5–3.7 billion years, but four gave ages averaging 4.1 billion years. Seventeen zircons from a nearby area have ages exceeding 4 billion years, and one gave the venerable age of 4.28 ± 0.06 billion years. But even older ages were to come.

Further study of the Yilgarn zircons, reported in January 2001, uncovered one with an age of 4.404 ± 8 billion years! This zircon somehow formed a mere 100 million years after the earth solidified. Since zircon occurs only in rocks of the upper continental crust, its presence 4.4 billion years ago shows that continental crust already existed at that time. To make matters even more intriguing, the isotopic composition of oxygen in the zircon shows that the magma from which it crystallized interacted with liquid water at low temperatures. This tiny grain of zircon has waited 4.4 billion years to reveal the secrets of the early earth and to tell us that continents and oceans may have gotten a much earlier start than anyone could have guessed. Two hundred years ago, James Hutton could find "no vestige of a beginning." The phenomenal discoveries of the geochronologists and their SHRIMP have moved us back 97 percent of the way to the beginning.

Patterson Confirmed

Prior to the discovery of the ancient zircons, in order to determine the true age of the earth it was necessary to have a model of lead evolution and to make at least one assumption. The best method used galena, the same mineral used by the lead isotope pioneers. Sometimes galena interlayers with marine sedimentary rocks like limestone, where, we believe, it arrived through the following process. The lead in the source region from which the galenas come—the upper mantle or lower crust—started out with the primordial lead of the solar system, but over time accumulated lead from the decay of uranium and thorium. Periodically, gases and hot fluids extracted metals, including lead, from the source regions and sent them upward along volcanic conduits to the surface. There, galena and other minerals precipitated in seawater and became interbedded with ocean floor sediments. Since galena contains no uranium, once formed, its lead isotope composition never changes. This process, repeated throughout the history of the earth, brought to the surface a series of galenas of different ages, each a fossil of the lead that existed in the mantle at the time galena formed. A plot of the lead isotope composition of these galenas would track the evolution of lead in the upper mantle and allow one to extrapolate backwards to the age of the earth.

The method has two potential problems. First, by definition, when the upper mantle or lower crust loses lead, it is no longer a closed system. Yet because the volume of lead extracted over time has been infinitesimally small compared to the total amount of lead, in practice the closed system requirement is not significantly violated. Second, some lead ores have been remobilized and deposited a second time, moving them off the line of lead evolution. But geologists can usually recognize such second-stage galenas.

The best independent age of the earth, based on the fewest assumptions, applies this backtracking method to four galenas from three continents known to be at least 2.7 billion years old and that geologists assume to have had the history just described. If each started out with the primordial lead isotope ratios of the Canyon Diablo meteorite, these leads define a line of evolution that traces back to an age for the earth of 4.54 billion years.

Patterson was right: many strands come together to show that the earth, the moon, and the meteorites all began 4.5 billion years ago with the same primordial isotopic abundances.

PART II

DRIFT

5

A Science Without a Theory

A science that hesitates to forget its founders is lost.

—ALFRED NORTH WHITEHEAD

AT ITS ANNUAL MEETING in New York City in 1926, the American Association of Petroleum Geologists (AAPG) convened a symposium on continental drift. Alfred Lothar Wegener (1880–1930), a German meteorologist and polar explorer, attended. A decade before, he had proposed the theory of continental drift in its modern form. Eminent geologists from America and abroad contributed: Bailey Willis of Stanford; Rollin T. Chamberlin, son of Thomas; John Joly; Charles Schuchert and Chester Longwell of Yale. After a decade in which to consider Wegener's theory, the response was largely negative. Rollin Chamberlin summed up the reaction, in words far more prophetic than any of the attendees could have realized: "If we are to believe Wegener's hypothesis we must forget everything which has been learned in the last 70 years and start all over again."

The assembled luminaries of geology did not reject continental drift because they already had satisfactory solutions to their most important problems. Just the opposite. They lacked a global theory of earth behavior and the answers to many fundamental questions: Why does the earth's surface divide into continents and ocean basins? Are the two regions essentially the same, but for their altitude? If so, could a continent be transformed into an ocean basin by lowering it, and conversely? Are continents and ocean basins permanent? What creates mountain ranges? Since erosion wears mountains down, why do they still stand high? Why do mountain ranges tend to be located on the edges of continents? Why has the sea invaded the land and retreated over and over again? Why do rocks formed in cold climates—glacial deposits, for example—today often reside at the equator, while rocks deposited in warm climates now occur near the poles? Why do continents separated by thousands of miles of ocean water contain the same nonmarine fossils? With such questions unanswered, the field of geology was paralyzed.

Just when geology was most in need of a global theory, Alfred Wegener came along to provide it. But, as Schopenhauer observed, the one who reveals the true path may be ridiculed, or scorned. In 1930, alone on the Greenland ice, Wegener died, a prophet without honor in the land of geology.

By the end of the nineteenth century, though a theory to tie the facts of geology together was lacking, many of the facts were well established. Geology as a descriptive science had made great strides, particularly in Europe and North America. An array of geological observations stood ready to be woven together into a theory of earth behavior. European geologists were more likely do so than their American colleagues, who still had a continent's geology to explore. Eduard Suess (1831–1914), an Austrian, theorized that the earth resembled a desiccating apple whose rind wrinkles as its interior shrivels. According to Suess, as Kelvin's cooling earth contracted, its skin—the rocks of the surface—crumpled upon itself to form mountain ranges. Huge sections of crust had collapsed into the interior, creating ocean basins and leaving the adjacent regions to stand as lofty continents. As the shriveling continued, new sections collapsed, forming new ocean basins and new continents beside them. Suess published his theory in four volumes entitled *The Face of the Earth.* In 1901, the president of the Bavarian Royal Academy of Sciences declared that "Suess has secured almost general recognition for the contraction theory. The work of Suess marks the end of the first day, when there was light." The history of science tells us that such unrestrained praise arrives just when a theory is about to collapse.

Suess's contracting earth theory was born in the geology of the Alps; Alpine geology struck the first blow. As geologists began mentally to unravel the intricate Alpine folds, they could see that some force had folded the rocks, then folded the folds. Some folds had been pushed over on their sides, showing that horizontal forces, rather than the vertical ones of Suess, had created them. When the unraveling was complete, it revealed that folding had compressed the rocks several times over: an Alpine belt that is now, say, 100 kilometers wide, before folding might have been 500 kilometers wide. If this degree of shortening were extrapolated to mountain ranges all around the earth, an enormous amount of shrinkage would be required—far more than any apple ever experienced.

The second blow to the contracting earth theory came from another mountain belt, the Himalayas. Sir George Everest (1790–1866), Surveyor-General of India, obtained a different result when he calculated the distance between two towns using ground-based surveying equipment than when he used astronomical observations. Everest reasoned that the great mass of the Himalayas might have attracted his plumb bob enough to throw off the ground calculations. But when he checked, he found just the opposite: the plumb bob swung away from the mountains, as if they were hollow. This led to the idea that a deficiency of mass beneath high mountain ranges offsets their obviously substantial bulk. If mountains are made of rocks less dense than the surrounding plains, and if mountains have deep roots that extend as far down into the earth as their peaks reach up, a plumb bob would swing away. Mountains may be like icebergs, extending downward in proportion to their height.

The concept of mountains floating in the substratum led to the theory of isostasy (equal standing), in which each region of the earth's crust rides in the denser mantle below. The postglacial rebounding of Scandinavia provides one crucial piece of evidence in favor of isostasy. Within recorded history—even within a single lifetime—the coastlines of Scandinavia have risen slowly out of the sea. Isostasy explains that the glacial ice that covered Scandinavia until only twelve thousand years ago depressed the crust, which sank into the substratum, like a loaded ship. When the glaciers melted, Scandinavia rose.

Isostasy appeared to explain another puzzle. When the folds and faults in mountain belts such as the Alps and Appalachians are unraveled, not only are the original rocks of great lateral extent, they are also miles thick. Yet when sedimentologists examine the sedimentary rock layers of mountain belts in detail, they find that each layer accumulated in shallow water. But how could deposition in shallow water produce a thick stack of rocks? A shallow trough, receiving sediment, sooner or later ought simply to fill up. Some thought that the weight of the accumulating sediments, like the weight of ice covering Scandinavia, caused the mantle beneath to flow out of the way, allowing the floor of the trough to subside just enough to make room for the next layer of sediment, and so on.

Geologists called these subsiding troughs geosynclines (a syncline is a downfold in layered rocks in the shape of a U). Exactly what force could transform a wedge of sediment miles deep into a mountain range miles high was a mystery. But geologists had to accept the evidence of their senses: where today we find mountain ranges, thick wedges of sediment often once existed. By showing that the less dense continents cannot sink to become ocean basins, isostasy delivered the second blow to Suess's theory of a contracting earth.

The third blow came from the discovery that radioactive decay releases heat. The earth need not be cooling, contracting, and running down, as Suess thought. Radioactive heat may have brought the earth to a stable temperature, or even have caused it to warm and expand. As it had to Kelvin's time scale, radioactivity delivered the *coup de grâce* to Suess's contracting earth.

If continents cannot become oceans, and vice versa, then each must be permanent. James Dwight Dana (1813–1895), Silliman Professor of Geology at Yale, proposed "permanentism" in the late nineteenth century. Continents may grow larger as some unknown process converts marginal geosynclines into mountains, but they can no more sink into the mantle than an iceberg can sink into the sea. Dana argued that the continents and ocean basins are primordial, permanent features of the earth's surface.

As the contracting earth theory waned, the concept of permanent continents and oceans took its place. One of the most influential proponents of permanentism in the new century was Bailey Willis (1857–1949), whose motto was, "Once a continent, always a continent; once an ocean, always an ocean." In 1910, Willis went Dana one better, saying, "the great ocean basins are permanent features of the earth's surface and they have existed, where they are now, with moderate changes of outline, since the waters first gathered." Between Dana's slogan and Willis's declaration lay a distinction with a difference. By adding the four little words, "where they are now," Willis changed the argument fundamentally. He claimed that not only are continents permanent in the sense that once formed, they do not disappear; they are permanently anchored at the same spot on the globe.

In spite of Willis's pronouncement, by the teens of the twentieth century not only had the contracting earth theory faltered, so had permanence. Contraction conflicted with Alpine geology, isostasy, and radioactivity. Permanent continents, with vast permanent oceans in between, conflicted with the abundant evidence from paleontology that the flora and fauna of widely separated continents had often been identical. To salvage permanentism, paleontologists conjured up elongated land bridges, which they said had once snaked across the oceans, connecting continents and providing routes across which itinerant organisms could migrate. To have stood safely above the waters, land bridges would have to have been made of less dense, continental rock. But, given isostasy, land bridges could not then have sunk into the denser mantle below. If the land bridges had not sunk, where had they gone? The story reminded some of the legend of Atlantis. Perhaps the missing land bridges might have provided not only the migration route for countless species, but Lebensraum for a vanished civilization.

As the contracting earth and permanence theories stumbled, geology had no replacement. If the earth were not contracting, there was no motor to drive geological processes. If the continents had always been in the same place, there was no way to explain the fossil evidence. Geosynclinal theory was little help. As Dana put it, referring to the mysterious forces that were supposed to have converted sediment wedges into mountains, geosynclines were part of a theory for the elevation of mountains with the elevation of mountains left out. As the nineteenth century closed, "The best geological ideas . . . had been refuted, and geologists found themselves with no viable account of their most basic and agreed-upon observational phenomena." A new synthesis was needed, one that tied together isostasy, radioactivity, mountain building, paleontology, and the geology of different continents.

6

The Dream of a Great Poet

If at first the idea is not absurd, then there is no hope for it.

—ALBERT EINSTEIN

IN 1914, THE GERMAN ARMY drafted thirty-four-year-old Alfred Wegener and sent him to a field regiment advancing into Belgium. He was shot twice, declared unfit for active service, and sent to a hospital to recuperate. That Wegener would be wounded in battle was no surprise, for he was a contradiction: a scholar who loved adventure. As young men, he and his brother had carried out what was then the longest balloon flight. Later, Wegener wintered over in Greenland with the Danish National Expedition. He returned to a lectureship at the University of Marburg, specializing in meteorology and paleoclimatology. Just before the outbreak of World War I, Wegener crossed the Greenland ice cap from east to west.

His recuperation from wartime wounds gave Wegener time to pursue an interest that he said developed by accident:

> The first concept of continental drift first came to me as far back as 1910, when considering the map of the world, under the direct impression produced by the congruence of the coastlines on either side of the Atlantic. At first I did not pay attention to the idea because I regarded it as improbable. In the fall of 1911, I came quite accidentally on a synoptic report in which I learned for the first time of paleontological evidence for a former land bridge between Brazil and Africa. As a result I undertook a cursory examination of relevant research in the fields of geology and paleontology and this provided immediately such weighty corroboration that a conviction of the fundamental soundness of the idea took root in my mind.

In his fine book *The Ocean of Truth,* William Menard aptly describes Wegener's state of mind:

> Wegener was a very trusting man. When he read about isostasy he believed that the geophysicists were right and that continents cannot be changed into ocean basins. When he read about fossils he believed that the paleontologists were right and that land species in Africa had had a way to reach South America. What is truly remarkable is that, unlike any contemporary, he managed to believe these things simultaneously.

Wegener presented his theory in a book entitled *Die Entstehung der Kontinente and Ozeane: (The Origin of Continents and Oceans).* The book first appeared in German in 1915, then in subsequent editions in several languages, the fourth and last in 1929.

Wegener's recollection of how he conceived the theory of drift in part explains its poor reception. He did no research, but got the idea while examining a map of the world and noticing how the coastlines of South America and Africa fit together. A year later, he came accidentally upon a report of the fossil evidence for a South Atlantic land bridge. Wegener thus presents himself as having discovered by accident the unifying theory of geology that had eluded all others.

Wegener began his book by reviewing and summarily dismissing the theories of contraction, permanence, and subsiding land bridges—isostasy falsified all three, he said. The only alternative interpretation of earth history is that the continents have drifted. Near the end of the Permian period, 250 million years ago, Wegener said, a supercontinent called Pangaea comprised all the present continents. In the Mesozoic, Pangaea began to subdivide, first in the southern hemisphere and then in the northern. India moved north and crashed into Asia, creating the Himalayas. The westward drift of North and South America against the resistance of the Pacific Ocean floor compressed their leading edges, thrusting up the Andean range that stretches the length of North and South America. Madagascar and New Zealand calved from their parent continents. "Smaller portions of blocks" flaked off and became the Indonesian chain, the Antilles, and the granitic Falklands and other islands of the Scotia Arc. Wegener's theory explained the origin of mountains, accommodated isostasy, and resolved the distribution of the flora and fauna of the past, and it did so in a way that anyone could understand at a glance.

Aware that it would take far more than a glance to convince geologists, Wegener presented a mass of supporting evidence. He went well beyond the jigsaw puzzle fit of South America and Africa to compare the geology of the two continents: "It is just as if we were to refit the torn pieces of a newspaper by matching their edges and then check whether the lines of print run smoothly across. If they do, then there is nothing left but to conclude that the pieces were in fact joined in this way." The lines of newsprint were the geological features of the continents on either side of the Atlantic.

Of course, the most convincing evidence would come from showing that the continents drift today. Wegener believed that Greenland had split from Scandinavia within the last hundred thousand years and had moved rapidly to the west. Precise measurements of the longitude of Greenland over time, therefore, might reveal this motion and confirm drift. In the final version of his book, he drew attention to measurements made in 1927 which indicated that Greenland was drifting at 36 meters per year. Only Wegener seems to have taken these spurious measurements seriously.

From the weak argument of longitude, Wegener moved on to the geophysical evidence, focusing particularly on isostasy, which "depends on the idea that the crustal underlayer has a certain degree of fluidity." And then the seemingly irrefutable key point: "If . . . continental blocks really do float on a fluid, there is clearly no reason why their movement should only occur vertically and not also horizontally, provided only that there are forces in existence which tend to displace continents, and that these forces last for geological epochs."

By the 1929 edition of his book, Wegener had the advantage of being able to read the lines of newsprint with a superb magnifying glass. In 1927, South African Alexander du Toit, who had been called the "world's greatest field geologist," made an extensive field comparison of South American and African geology. He found so many similarities that to summarize them took seven pages. Du Toit's most striking conclusion was that the geology of a given point on the African coast more closely resembles the geology of the opposing point on the South American coast than either resembles the geology immediately within its own continent. In other words, the coastal geology of South America is more like that of Africa than it is like that of the rest of South America, and vice versa. Wegener cited du Toit's conclusions and went on to compare the geology of the continents bordering the North Atlantic, where he also found many similarities.

Wegener's chapter on paleontology used the faunal and floral identities to establish that the continents had once been joined. One particularly convincing piece of evidence was the small Permian reptile,

Mesosaurus, found only in southeastern Brazil and southwest Africa, and in identical rocks. Since *Mesosaurus* could not have swum the Atlantic, one had to choose either land bridges or drift. The other frequently cited evidence was the fossil plant *Glossopteris,* found in all the southern hemisphere continents. From fossils, Wegener moved to consider living organisms. How to explain the remarkable similarity of modern earthworm genera on different continents? They can neither swim nor fly; it seems far-fetched to imagine worms making a transatlantic crossing by scooting along the length of a 6,000-kilometer land bridge. And what of the peculiar Australian marsupials and monotremes (the platypus, for example) that have no close relatives in the rest of Asia, but whose South American cousins closely resemble them, down to their parasites?

Wegener was neither a geologist nor a paleontologist; in those fields, he had to rely on the work of others. But in meteorology and paleoclimatology he was expert, having written an important book with his father-in-law. In these disciplines, Wegener found evidence even harder to explain without continental drift than *Mesosaurus, Glossopteris,* and perambulating platypuses and earthworms.

Geologists interpret certain rock deposits as evidence of past climates. Glacial tillites indicate cold conditions; coal, high humidity; desert sandstones, aridity; salt and gypsum, warmth and evaporation. Wegener showed that past climates differed markedly from present ones: Spitzbergen is now in the cold North Atlantic but in the late Paleozoic had a subtropical climate. At that time, ice caps covered South Africa and Australia. The obvious explanation—a wholesale warming or cooling of the earth's climate—fails because some of the climate changes occurred at the same time: while one region was growing colder, another was growing warmer.

The most striking support of drift from past climates came from late Paleozoic glacial deposits found in all the southern hemisphere continents and even in the far-offshore Falkland Islands. Geologists could tell from glacial striations that the ice that invaded South America had come from the southeast, where now resides only Atlantic Ocean water. But while glaciers covered South America, the northern hemisphere continents were ice-free. Once again, a worldwide change in climate could not explain the evidence. Wegener showed how simply drift could resolve the conundrum: merely assume that the southern hemisphere continents once clustered around the South Pole. QED.

At this point in the final edition of his book, having reviewed a mountain of evidence confirming drift, Wegener could hold back no longer:

> We shall refrain here from citing the literature in support of our statements. The obvious needs no backing by outside opinion, and the willfully blind cannot be helped by any means. As far as we are concerned, it is not now a question of whether the continental blocks have moved; doubt is no longer possible.

Even those who would not see could reasonably ask what might have caused the continents to drift. Wegener did not know, but he took an educated guess. The earth bulges slightly at the equator, causing an object on the surface to experience a force away from the poles. Wegener thought that this "pole-fleeing" force might cause continents floating isostatically in a fluid mantle to drift slowly toward the equator. Tidal friction and other forces might push the continents toward the west, where the resistance of the ocean floor would thrust up the Andean chain. But Wegener seems to have had no illusion that these weak and speculative forces would satisfy anyone—they did not satisfy even him. He noted that science should proceed empirically, with the amassing of evidence without regard to whether a theory was yet available to explain the evidence. The "Newton of drift theory has not yet appeared," said Wegener, but "his absence need cause no anxiety; the theory is still young and still often treated with suspicion."

Wegener's purported mechanisms of drift were far too weak to move continents and served merely to call attention to their weakness. In hindsight, he would have done better to leave the cause of drift to others, as his words above suggest he was ready to do. But Wegener was an honest, if naive, man. He did his best to come up with a mechanism for drift, failed, and admitted it. In so doing, he handed his opponents a weapon.

In 1929, the year of the final edition of Wegener's book, one who would come as close to a Newton as geology would find in the twentieth century had already deduced the mechanism of drift and had begun to speak and write about it. Yet for three decades, geologists ignored his theory. Then they endorsed it and gave the credit to one of their own.

7

The Rejection of Drift

To be uncertain is to be uncomfortable but to be certain is to be ridiculous.

—Goethe

Made by Satan

The reaction to Wegener's 1915 book took a few years to set in. When it did, the response was profoundly negative. Wegener's methods drew as much of the initial criticism as his science. "He is not seeking truth; he is advocating a cause, and is blind to every fact and argument that tells against it," said one critic." "Science has developed by the painstaking comparison of observations and, through close induction, by taking one short step backward to their cause; not by first guessing at the cause and then deducing the phenomena," said another.

By the time of the 1926 meeting of the American Association of Petroleum Geologists (AAPG), opinions had hardened. The convenor, the magnificently named Dutch scientist W. J. A. M. van Waterschoot van der Gracht, tried to maintain balance, but he was the exception. The other presenters, especially those from North America, while professing objectivity, had made up their minds. Bailey Willis wrote: "After considering the theory of continental drift with avowed impartiality, the author concludes . . . that it should be rejected." Rejection was mandatory for Willis because the theory "sprang from a similarity of form (coast lines of Africa and South America) which . . . drift would have destroyed . . . by faulting. If similarity of outline be a fact, then there . . . can have been no movement of one continent away from the other." Willis turned the jigsaw puzzle fit of Africa and South America inside out, claiming that the match was too good to be true. If the two continents were once joined, he argued, the forces that tore them apart would have destroyed the fit. Therefore, following Willis, the better the fit between two continents, the less likely they are to have been connected. Such purported logic, unrefuted, from one of the most eminent geologists of his day, tells us much about the state of geology at the end of the first quarter of the twentieth century.

Rollin Chamberlin wondered (attributing the question to "other groups of the profession"):

> Can we call geology a science when there exists such difference of opinion on fundamental matters as to make it possible for such a theory . . . to run wild? Wegener's hypothesis in general is of the footloose type, in that it takes considerable liberty with our globe, and is less bound by restrictions or tied down by awkward, ugly facts than most of its rival theories. Its appeal seems to lie in the fact that it plays a game in which there are few restrictive rules and no sharply drawn code of conduct.

E. W. Berry, professor of paleontology at Johns Hopkins University, asked that it be "borne in mind that imagination by itself has never widened the bounds of knowledge." He too objected to "the author's method [which] is not scientific, but takes the familiar course of an initial idea, a selective search through the literature for corroborative evidence, ignoring most of the facts that are opposed to the idea, and ending in a state of auto-intoxication in which the subjective idea comes to be considered as an objective fact." (Naomi Oreskes notes that in the 1920s, "auto-intoxication" was a "euphemism for constipation.") Charles Schuchert (1858–1942), professor of geology at Yale and the dean of American paleontologists and historical geologists, also criticized Wegener's methods:

> He generalizes too easily from other generalizations, and . . . pays little or no attention to historical geology or to the time of the making of the structural and biologic phenomena discussed. Facts are facts, and it is from facts that we make our generalizations, from the little to the great, and it is wrong for a stranger to the facts he handles to generalize from them to other generalizations.

Schuchert provided the longest and most influential criticism among the symposium papers. To understand the condition of geology in the 1920s—and indeed, through to the 1960s—no written statement is as important. With the advantage of hindsight, Schuchert's article most clearly reveals the box into which geology had placed itself and why Wegener was unable to receive a fair hearing.

Schuchert began his paper by noting that a friend had remarked that the similarity of the coastlines of Africa and South America must have been "made by Satan" to vex geologists. Where Willis theorized, Schuchert experimented. He constructed an 8-inch globe of the earth, placed Plasticine tracings over the continents bordering the Atlantic, and slid them around to see how well they fit. He found that a close fit in one spot produced either a gap or an overlap in another, sometimes amounting to an error of 1,500 miles. Wegener had thus "taken extraordinary liberties with the earth's rigid crust," said Schuchert, concluding that the large misfit between the continents falsified Wegener's theory. A few pages later, Schuchert asks: "Are we to believe, with Wegener, that shore lines and shelf seas have remained constant in shape, position, and contour during 120 million years?" Here Schuchert joins Willis in arguing that geologic processes since drift began would have destroyed whatever fit there had been just after the original split of the continents. The fit was poor. But if it was not poor, it was too good to be true.

Schuchert acknowledged some important congruities in the geological newsprint: "It can be truthfully said that Wegener's hypothesis has its greatest support in the well known geologic similarities on the two sides of the Atlantic, as shown in strikes and times of mountain-making, in formational and faunal sequences, and in petrography." But whatever the resemblances, they were not enough: "slight similarities should be striking identities."

Schuchert wrote that he was "iconoclastic toward the Wegener hypothesis as a whole," yet was "wholly open minded toward the idea that the continents may have moved slowly, very slowly indeed, laterally, and differently at different times." He wavered enough to say: "When one turns to the Alps and is told by the best of authorities that their present width of some 150 miles was originally 500 and perhaps 625 . . . he begins to remember the statement of Galileo in regard to the earth: 'And yet it does move.'"

As Oreskes has deduced, Schuchert had not been reading Galileo. Instead, he had no choice but to read and consider a new book entitled *Our Mobile Earth,* by a respected Harvard professor, Reginald Daly. Daly's book took as its motto Galileo's famous sotto voce parting shot at his Vatican inquisitors. Daly was one of the few American geologists willing even to consider drift. He was also one of the few to have read Wegener in the original German and to have spent time in the field in South Africa with du Toit. Geologists had to take Daly seriously. (In the biographical memoir for Daly published in 1960, his successor at Harvard, Frances Birch, says not a word of continental drift, even though it had been a focus of Daly's work for decades. Birch was a follower of the Cambridge geophysicist Sir Harold Jeffreys, and like him a bitter opponent of drift.)

Instead of retaining drift among his working hypotheses, Schuchert preferred land bridges, in which he had "long been a believer . . . all through the Paleozoic and Mesozoic across the Atlantic from Brazil to Africa." While faulting Wegener for an overactive imagination, his critics endorsed land bridges, which were entirely imaginary.

When people take a stand that hindsight shows to have been illogical, we may suspect that ideology motivated them. In the concluding section of his AAPG paper, Schuchert confirms the suspicion: "We are on safe ground only so long as we follow the teachings of the law of uniformity in the operation of nature's laws." To accept drift would have required him to abandon uniformitarianism; land bridges kept it afloat.

Beyond his seminal work on fossils, Schuchert's main professional contributions were maps that showed the arrangement of land and sea at various times in the past. Implicit in the creation of such maps are continents and oceans always located where we find them today. If one does not know where on the globe North America lay at some point in the past, one cannot make a map that shows its position. If the continents could have been anywhere at any time, how could one use paleontology to understand broad earth history?

After all, why should one give up the time-tested rule of uniformitarianism in favor of the ideas of a German meteorologist who refused to play by the rules? Wegener had discovered his theory by accident and done no original work, had written it up while recovering from wounds suffered defending the Axis, had failed to consider alternative hypotheses, and had labeled his opponents "willfully blind" to "obvious" conclusions. To add insult to injury, in the final edition of his book, just after the AAPG Symposium had rejected drift, Wegener declared that "doubt is no longer possible."

Fairy Tales

As an outgrowth of the AAPG Symposium, Schuchert and Willis began an extensive correspondence with each other and separately with Arthur Holmes.* Holmes had shown a deep interest not only in the age of the earth but in all the broad questions of geology. He thought that radioactive heat might drive mountain building, and, possibly, move continents.

Schuchert wrote to Holmes that "the South Atlantic land bridge is a fact for Paleozoic and much of Mesozoic time. We must have a way of getting rid of it in early Cretaceous time." But Holmes saw "no alternative at all to continental drift."

In another letter, written in 1929, Schuchert revealed his dilemma: "In spite of all the . . . American criticism about Gondwana, and the fact that I *want* to give up the theory, so far I have not been able to explain away what I know of the marine faunas." (Gondwana being the hypothetical assemblage of southern hemisphere continents prior to drift.) In other words, without a previous connection between the continents, Schuchert could not explain the evidence of paleontology. The problem was how to get rid of land bridges once species had crossed. In his symposium article, Schuchert put this problem off to the future, expressing confidence that "the geophysicists will in time find the way in which [the breaking down of land bridges] was accomplished." Schuchert and his colleagues did not extend to Wegener the same privilege of being able to rely on the work of future geophysicists to solve problems, such as discovering the mechanism of drift. Instead, Schuchert's newfound colleague Bailey Willis made the work of the geophysicists unnecessary. Willis came not to bury land bridges, but to raise them.

---

*Oreskes (1999) reviews the private correspondence of these men.

Willis and Schuchert were both born before the Civil War and lived well into, and in Willis's case beyond, World War II. In 1932, they published back-to-back papers in the prestigious *Bulletin of the Geological Society of America.* Willis was then seventy-three, Schuchert seventy-two. The two would have headed any short list of the Grand Old Men of their profession.

At the outset, however, they were far apart. Schuchert thought that land bridges were a fact. Willis believed that continents and oceans were permanent and that less dense land bridges could not sink into the oceanic crust and mantle. How to resolve the dilemma? Willis came up with a new idea that preserved the essential features of both philosophies. Land bridges had existed, but instead of being made of continental rock, which would float stubbornly, they were made of dense oceanic basalt, thrust up from the ocean floor. Basalt, being dense, would sink again when convenient. These "uptrusions" formed long, slender connections that Willis, ever fond of a neologism, called "isthmian links." Willis's theory was even uniformitarian: basaltic islands are common; Central America is an isthmus that links two continents.

Schuchert's 1932 paper, entitled "Gondwana Land Bridges," summarized the evidence from paleontology and paleogeography that he had assembled over his long career. His maps of land bridges appear as realistic as if they referred to known land masses. Looking at his maps today, it is hard not to place them beside sketches of the canals of Mars, or the maps of alien lands in a science fiction novel. That no evidence existed for land bridges did not temper Schuchert's confidence. The U.S. Navy had begun to sound the oceans and Schuchert believed that they or someone would find his foundered land bridges.

The pair of papers by the distinguished septuagenarians, published in a distinguished American journal and expressed with the utmost assurance, "effectively marked the end of active debate over continental drift in the United States." As if to drive the final nail in the coffin of drift, in 1944, when he was eighty-five, Willis wrote a paper entitled

"Continental Drift, ein Märchen (a Fairytale)." Against drift was "conclusive negative evidence" that should cause it to "be placed in the discard, since further discussion of it merely incumbers [sic] the literature and befogs the minds of fellow students." Willis wanted to banish drift not just from the list of working hypotheses but from the very curriculum of geology.

In the same year in which the Fairy Tale paper appeared, Bailey Willis received the Penrose Medal of the Geological Society of America, "awarded in recognition of eminent research in pure geology, for outstanding original contributions or achievements that mark a major advance in the science of geology." The message to geologists of the 1940s and 1950s could not have been more clear: Oppose drift, indeed, treat it first as a fairy tale and then as subversive, and reach the pinnacle of your profession. While the converse might not have been true—Harvard's Daly, who supported drift, was an obvious exception—that might merely prove the rule. What young geologist would take the risk?

8

A Plausible Mechanism

Anything that has happened, can.

—Marshall Kay

For Lack of a Cause

The generation that gathered at the 1926 AAPG Symposium rejected drift for unscientific and contradictory reasons. As the years passed, successive generations of inquisitive students needed a more flattering explanation of why the brightest and best of their predecessors had abandoned drift. No one could fail to see that considerable evidence supported the theory, some of it obvious from a glance at a map of the world, some painstakingly assembled by du Toit and others. Even more considerable evidence must count against drift. What was it?

No one could claim that another theory had challenged and defeated drift. There was no other theory. Land bridges did not account for global geology—paleontologists conjured them up merely to explain the identity of fossil species on widely separated continents. On the other hand, given the lack of knowledge about the ocean floor in the

1920s, no one could prove that land bridges had not once existed. For those who preferred to doubt drift, the bridges provided a convenient, if now submerged, way out. Geosynclinal theory served little better; it was still a theory for the origin of mountains with the origin of mountains left out. Geologists could tell that the forces that had compressed the shallow-water sediments of the Appalachian geosyncline had come from the southeast. But when students inquired as to the source of these mysterious pressures, from where now only ocean breezes blow, professors could only shrug, in effect adopting the stance of Schuchert: future work may supply the answer.

If a better theory did not replace drift, why did geologists reject it? Successive generations required an answer that was scientific and independent of Wegener's allegedly unscientific style. Wegener handed his critics one reason, and it served until the 1960s: his puny pole-fleeing and westward-drifting forces. Thus entered the lore of the profession the notion that geologists rejected drift for "lack of a mechanism."

But this claim was inconsistent, if not disingenuous. Every geologist who rejected drift believed that great continental ice sheets had several times advanced and retreated over the northern hemisphere, yet no one knew why. The drastic horizontal shortening in the Alps was plain to see, yet no one knew why. People accepted electricity, as well as the earth's magnetism and gravity, before anyone could explain them. Scientists have admitted any number of observational phenomena before they knew their cause. That Wegener could not come up with a plausible mechanism for drift was a convenient shibboleth, a way for one generation to explain to the next the reasons for the rejection of continental drift.

Given the longevity of the myth of Wegener's "lack of a mechanism," it is surprising to find that, by the time the AAPG Symposium book appeared in print, a plausible mechanism for drift had been proposed. Schuchert, Willis, and other eminent geologists became well aware of the proposed cause and corresponded about it.

The mechanism was convection in the earth's mantle. In a heated room, for example, air convects constantly. As warmer air expands, its molecules separate, making it lighter. The lighter air rises to the ceiling, where it cools and begins to sink, starting the process over again. The same thing happens in a pot of boiling water. To convect, a material must be able to move physically, as can a gas or liquid. If the mantle convects, over the long run it must behave as a fluid, as isostasy suggests it does.

But how can we be sure today that geologists of the 1920s and 1930s should at least have admitted convection to their list of working hypotheses? Because, four decades later, a new generation enthusiastically adopted convection as the driver of drifting continents and spreading plates.

Radioactive Heat

John Joly was one of the last holdouts against an earth much older than the 100 million years that his salt clock had revealed to him. He tried to show that the rate of radioactive decay varies over time, but could not. Yet, in another way, Joly's interest in radioactivity proved prescient. Indeed, it would not be much of a stretch to credit him with the original conception of the mechanism of drift—mantle convection—though the idea goes back even earlier. In any event, the one who did more to develop the idea of convection as the cause of drift than anyone else was Arthur Holmes, who deserves priority.

Joly was quick to understand that radioactivity might provide not only the clock to measure the earth's age but, through its release of heat, the energy to drive geological processes. In his *Radioactivity and Geology,* published in 1909, Joly named radioactive decay as the vera causa of geological dynamics. He described how radioactive heat builds up until rocks melt. The resulting hot, expanded magma uplifts the crust, pushes up mountains, and erupts on the surface.

Joly thought that radioactive heating would cause the hotter lower mantle to change places with the cooler upper mantle. Such overturning might have happened repeatedly in earth history and might be the immediate driver of large-scale earth movements. It was but a short step to conjecture that the churning of a fluid mantle would cause the "continental masses [to] slowly replace the oceans."

Joly submitted his ideas on radioactive heating and convection to the *Philosophical Magazine* in 1923, but the editors rejected the paper. Ever afterwards, Joly adopted a cautious stance toward drift. His chapter in the AAPG book was only one page long. In his last major publication, the 1930 edition of his book *The Surface-History of the Earth,* he defended drift and cited new evidence for it.

But Joly was not the only, nor indeed the most successful, geologist to take up the question of radioactivity in the early years of the twentieth century. The other was a much younger man, Arthur Holmes. He would prove to have the energy, the intellect, the self-confidence, and the staying power to develop fully the idea of convection as the driving mechanism of drift. Joly passed the torch of radioactive heating to Holmes, who used it to warm the mantle and move the continents.

Convection, 1929

By the 1920s, Arthur Holmes was the leading figure in determining the age of the earth, having long since made the transition from physics to geology. As a result of his work in Africa and Burma, he was widely experienced in geology; by 1930, he had written dozens of papers. No one was better prepared to apply physics to geology or better able to understand the implications of radioactive heating.

In 1915, when he was twenty-five years old, Holmes wrote two papers in a series called "Radioactivity and the Earth's Thermal History." The third followed a year later; then came a hiatus as Holmes published on more immediately practical geological topics, such as "The Geology of Concrete Aggregates." In 1925 and 1926, he returned to radioactive heat. Holmes agreed with Joly that the earth was unlikely to have achieved a constant temperature, otherwise, it would be dormant. Holmes did not think the earth had been warming, otherwise our planet would have grown uncomfortably hot. Cooling and Suessian contraction had failed. Joly's cycle of radioactive heating, uplift and magmatism, followed by cooling, with the process beginning again and repeating at intervals, was the best idea left.

In a talk to the Geological Society of Glasgow in 1928, published in 1929, Holmes directly linked radioactive heating and drift. His words and diagram (Figure 8.1) anticipated the plate tectonic revolution by four decades:

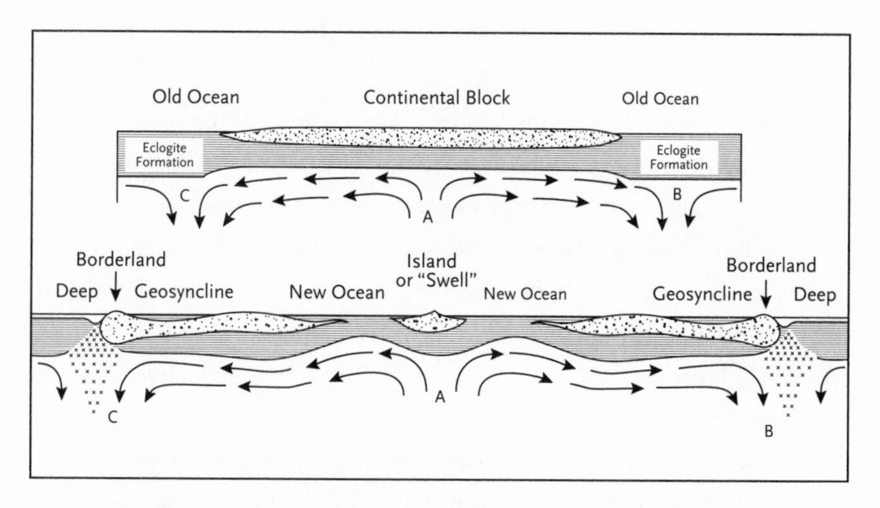

Figure 8.1 Arthur Holmes's Model of Sea Floor Spreading and Continental Drift (redrawn from Holmes, 1929).

Holmes agreed with Joly that hot mantle material at depth rises to the surface and flows to either side. As the material moves farther away from the upwelling cell, it cools, grows more dense, and sinks, beginning the process anew. Thus, radioactive heating generates giant cells of uprising and down-flowing mantle. A convection cell that rises and diverges beneath a continent pulls it apart, as shown in Holmes's diagram (Figure 8.1), and drags the sundered fragments, continents in their own right, to either side. In between, a new ocean basin forms. Since the lighter continental material cannot sink into the denser mantle, it piles up at the margins and forms mountains. The theory solves two old problems at once: "Mountain building would have been accomplished on the continental margins or on the sites of former geosynclines."

Holmes's logic was unrelenting: "We know that sites of considerable areas of the Atlantic and Indian Oceans were formerly occupied by continental masses and since these ancient lands are no longer there, we are driven to believe that their material has moved sideways." But sideways movement pointed to "continental drift on a scale of the same order of that advocated by Wegener." "Many geologists," Holmes noted in an understatement, "have hesitated to accept this consistent and straightforward reading of the rocks" because of the lack of a driving mechanism. And then on to his clincher: "Admitting that the continents have drifted, there seems no escape from the deduction that slow but overwhelmingly powerful currents must have been generated in the underworld, namely that convection currents may be set up in the lower layer as a result of differential heating by radioactivity." Drift has happened, therefore it can; convection is its driving force.

Holmes's theory, and its incorporation of earlier work on isostasy and drift, answered the major questions that had long vexed students of the earth. It explained why continents and ocean basins exist and persist. It explained the origin of geosynclines and showed how they become mountains. It explained the faunal evidence and the other geologic "newsprint" that Wegener and du Toit had assembled. It provided the driving mechanism for drift that critics said Wegener's theory lacked. Holmes's theory avoided one criticism leveled at Wegener's: he did not require that the continents ram their way through an unyielding ocean floor of rigid basalt. Instead, they ride passively atop a spreading convection cell, as if on a conveyor belt.

Holmes took Wegener's theory, with its many flaws but generally right concept, and improved it to the degree possible in the late 1920s. Certainly, Holmes got some things wrong. The eclogite (a high-pressure garnet pyroxene rock) shown on his diagram is not produced in the way he thought; mid-oceanic ridges are not fragments of continents (though such continental fragments do exist in the Falklands and Seychelles). But these are details. Four decades after Holmes's 1929 paper, geologists accepted convection as the cause of drift. We accept convection today, even though we still do not fully understand how it works.

Why did geologists not embrace convection and drift at the time Holmes linked them? We might think that in the days before faxes and ready airmail, North American geologists were unaware of Holmes's speech to the Geological Society of Glasgow and his subsequent paper. Not so. Naomi Oreskes has done students of geology the great favor of reviewing the elaborate and lengthy correspondence carried on between the leading geologists of America and Britain over drift in the 1920s and 1930s. Schuchert and Holmes were each fully aware of the latest ideas of the other. In 1927, Schuchert wrote Holmes that "the South Atlantic land bridge is a fact. . . . We must have a way of getting rid of it. . . ." He asked Holmes to consider the idea that the piling up of heavy basalt would cause a continent to sink, one possible way of disposing of the land bridges. "On the whole," Holmes replied: "I find a combination of Wegener's ideas with magmatic convection currents . . . to provide the energy. . . ." Later, responding to further entreaties from Schuchert, Holmes wrote: "It is impossible to get rid of lands that formerly occupied the sites of present oceans except by moving them sideways. I see no alternative to continental drift, and I have come to that conclusion from a position of strong prejudice against such processes."

In late 1930, Schuchert read Holmes's 1929 article in the *Transactions of the Geological Society of Glasgow.* Schuchert wrote a colleague: "It is along the lines laid down by Holmes that geologists will eventually solve the causes of reformative mountains and how they are made." Schuchert's wish to use the downturning convection cells to sink land bridges, once they had served their purpose, might explain his enthusiasm. But Holmes had clearly rejected the sinking of land bridges, much less entire continents. At about this time, Schuchert fell under the influence of Bailey Willis, for whom convection was "an impossibility." Willis admonished the apparently wavering Schuchert, "Pray don't get Wegener and Willis mixed up in your mind." Schuchert never again did.

For three decades after the Schuchert and Willis papers, with one major exception, no one wrote about drift. The exception was Alexander du Toit's book, *Our Wandering Continents,* published in 1937 and dedicated to Wegener. Du Toit marshaled all the impressive geologic evidence for drift; but North American geologists ignored his book as Wegener redux. After Willis's 1944 Fairy Tale article, in North America the theory of continental drift disappeared. To espouse drift was to run the risk of appearing silly, and, if one persisted, heretical. In American geology textbooks, journals, and courses from the end of World War II until the late 1960s, if mentioned at all, drift served only to provide an example of how geology had been tempted into error but had been saved by its allegiance to uniformitarianism. West of the mid-Atlantic ridge, drift was dead.

Willis and Holmes

Arthur Holmes spent the years of World War II at Durham University. With most young men—and few women studied geology—off to war, he had time to write a comprehensive textbook (and to think about lead isotopes). The first edition of his *Principles of Physical Geology* appeared in 1944; the second in 1965. His widow, Doris Reynolds, a fine geologist in her own right, completed a third edition in 1978.

Willis's Fairy Tale paper and Holmes's concluding chapter on continental drift could hardly have been more different. Willis attempted to stamp out continental drift once and for all through a combination of carefully selected negative evidence and ridicule, the ultimate form of which came from his title: drift was nothing more than a fairy tale.

But Willis could not get by with merely insulting the theory of continental drift; he had to provide reasons that would satisfy those who might not heed his warning that studying drift would befog their minds. To those who had followed Willis's work, his first objection sounded familiar: "It is a well established principle of mechanics that any floating object moving through air, water, or a viscous medium creates behind it a suction. . . ." Therefore, sections of South America, drifting west, would have been suctioned off and left behind in the Atlantic to the east, where we should find them. "Moreover, if such segments had been pulled off, the eastern shoreline of South America should not so closely resemble the other side of the supposedly original fracture. The close similarity of the two coasts presents us with the choice: either the laws of dynamics were suspended to preserve unaltered the rear outline of the drifting continent or the continent did not drift." In other words, eighteen years after the AAPG Symposium, the fit of the South Atlantic continents was still too good for Willis. After reminding readers that drift still lacked a driving mechanism, he attacked from a different direction.

> Fellow scientists who are not geologists cannot be expected to know that the geology upon which the protagonists of the Theory rest assumptions is as antiquated as pre-Curie physics. Wegener and his successors are disciples of Eduard Suess . . . a

> charming, genial German who never traveled far, but assembled the observations of others and from them constructed speculations regarding the face of the Earth. He gravely lacked critical faculty.

Thus, Willis sank so low as to find Wegener, a brave polar explorer who died alone on the Greenland ice, guilty by association with an alleged armchair geologist.

Willis then rhetorically sank Gondwanaland, the hypothetical southern hemisphere megacontinent, which "had no actual existence . . . [and] was the identical line of reasoning of the argument for Continental Drift. In Suess's imagination, [Gondwanaland] was a reality; but there is . . . no geologic fact to demonstrate [its] existence." He lauded Schuchert's land bridges, which are truly imaginary, as a worthy improvement over Gondwanaland, which today we know did once exist. Willis did acknowledge that he might have to adjust his theory of isthmian links to accommodate new evidence showing the similarity of fossils across the vast Pacific basin. He called upon "the emergence of lands which have now subsided beneath the waters" to solve the problem. With that, Willis's brief article ends.

In his 1944 text, Holmes took a different tack. He reiterated the ideas set forth in his 1929 paper about mantle convection and then moved on to drift. He set up the perennial dilemma of geology by asking whether the present continents and oceans have been "stable or otherwise during geological time." On the one hand, the absence of deep ocean sediments on the continents suggests stability. On the other, the fossil evidence indicates that the continents were once connected. But over the twentieth century, evidence had accumulated that "certain regions that were undoubtedly land long ago are now parts of the Atlantic Ocean. What happened to these vanished lands?" asked Holmes.

Holmes could see only two possible explanations of the broad facts of geology: drift; or a horizontal stretching out, after which isostasy had caused the thinned layers to sink. He reviewed the history of ideas about drift, noting that back in 1620, Francis Bacon had pointed out the resemblance between the coastlines of Africa and South America. Holmes assessed Wegener's ideas for the mechanism of drift as "hopelessly inadequate." He then wrote two sentences that showed just how different was his approach from that of his dogmatic American colleagues: "The really important point is not so much to disprove Wegener's particular views as to decide from the relevant evidence whether or not continental drift is a genuine variety of earth movement. Explanations may be safely left until we know with greater confidence what it is that needs to be explained." To Holmes, drift was not a fairy tale invented by an armchair German, but a genuine theory, far from proven but worthy of study. It belonged on a short list of multiple working hypotheses.

"The choice," he wrote, "evidently lies between accepting continental drift or postulating a giant land bridge across what is now 4,000 miles of ocean." Holmes believed that the evidence of the late Paleozoic glaciation of Gondwanaland would prove decisive. Geologists had found beds of glacial tillite "in India in 1857, in South Australia in 1859, in South Africa in 1870, and in Brazil in 1888," lands that now lie both above and below the present equator. It had become clear that the proposed great southern continent of Gondwanaland underwent glaciation at the same time the northern hemisphere "enjoyed mild or tropical climates." Since these marked and contradictory differences from the present climate patterns had taken place almost simultaneously, a worldwide cooling or warming trend could not be the cause. Glacial striations, and other evidence, showed that the ice had sometimes come from the open ocean. "With the continents in their present positions," Holmes wrote, "such a distribution of ice-sheets is hopelessly inexplicable." But not if "all the continents except Antarctica lay well to the south of the present positions . . . grouped together around the South Pole." As Wegener had proposed.

The only argument Holmes could find against drift was that "it merely exchanges one embarrassing problem for another—the difficulty of explaining how continental drift on so stupendous a scale could have been brought about." But, Holmes noted, if one rejects drift and the climatic evidence, one must then accept land bridges. "The continental drift solution has the advantage that it reduces two baffling problems to one," he wrote, referring to both the fossil and the climatic evidence.

Holmes proposed convection currents as "the sort of mechanism for which we are looking," and showed a slightly modified version of the diagram from his 1929 paper. He wrote that subcrustal currents ascending beneath a continental block "become sufficiently vigorous to drag the two halves of the original continent apart, with consequent mountain building in front where the currents are descending, and ocean floor development on the site of the gap, where the currents are ascending." The currents "would inevitably carry the continents along with them, provided that the enormous frontal resistance could be overcome." Holmes summed up: "During large-scale convective circulation the basaltic layer becomes a kind of endless travelling belt on the top of which a continent can be carried along, until it comes to rest (relative to the belt) when its advancing front reaches the place where the belt turns downwards and disappears into the earth."

Holmes ends his book with words of appropriate caution. To go beyond his outline of convection as a mechanism "would at present be unwise. Purely speculative ideas, specially invented to match the requirements, can have no scientific value until they acquire support from independent evidence." And finally, "Many generations of work, geological, experimental, and mathematical, may well be necessary before the hypothesis can be adequately tested." One reason Holmes may have thought that many generations would be required was that hardly anyone was working on the problem of drift. Wegener and du Toit were dead; in America, drift was a risible fairy tale. In 1944, isolated as he had been by a world war and military secrecy, Holmes could not have known that, on and under the world's oceans, a new breed of scientists was laying the foundation that would cause most geologists to accept the speculative idea of drift within only one generation.

9

Data from the Abyss

> The sea, washing the equator and the poles, offers its perilous aid, and the power and empire that follow it . . . "Beware of me," it says, "but if you can hold me, I am the key to all the lands."
>
> —RALPH WALDO EMERSON

On the Brink

In 1937, the oldest learned organization in the United States, the American Philosophical Society, founded under the impetus of Benjamin Franklin in 1743, and one of the newest, the American Geophysical Union, founded in 1919, held a symposium entitled "The Geophysical Exploration of the Ocean Bottom." Richard M. Field, professor of geology at Princeton, organized the meeting. At the AAPG Symposium in 1926, with no new data, the participants had little to do other than quote what each other thought about drift. For new research contributions, they had to resort to sliding Plasticine sheets around on an 8-inch globe. By the time of the 1937 conference, a shift that would prove titanic had begun: new data, and data of a different sort, were becoming available, and a new group of geologists, or rather geophysicists, was gathering and analyzing it. Increasingly, data and not personality would lead geology.

The new geophysical data included measurements of the strength of the earth's gravity field made from submarines. If isostasy were perfect, the surface features of the earth—the mountain ranges, plains, and ocean basins—would float in gravitational balance. Each region would rise or sink until it reached the level that accorded with its dimensions and density. The pull of gravity at the surface would then be the same everywhere. But if the crust moves laterally, the earth could not maintain isostatic balance and the force of gravity would be different at different places. Thus, the degree to which the crust preserves isostasy, as revealed by gravity measurements, provides a test of drift. But the differences would likely be slight at best and require new instruments to detect them.

The pioneer in measuring gravity at sea was a tall Dutchman named Felix Vening Meinesz, who in the early 1920s invented an instrument for measuring the pull of gravity for use in a submarine. Vening Meinesz folded himself into a primitive Dutch sub and sailed under the oceans, making hundreds of laborious gravity measurements. His most significant finding was that in the region of the deep-sea trenches, which lie on the perimeter of the ocean basins and produce the deepest ocean waters, the pull of gravity was much less than it should be if isostasy were preserved. Beneath the trenches, some inexorable force pulls the crust down. Vening Meinesz, following Holmes, proposed that the force was a down-welling convection current. On one voyage, a young American from Princeton named Harry Hess accompanied Vening Meinesz. Hess realized that by measuring gravity anomalies, geologists were studying the behavior of the earth, not as it was millions of years ago, but as it is today. They might be witnessing mountains forming.

Richard Field of Princeton opened the 1937 conference by setting up the decades-old dilemma of geology: on the one hand, geosynclines and permanence; on the other, continental drift. Field made his choice clear: the developing understanding from geophysics was "diametrically opposed to the earlier and simpler concept of the . . . Appalachian geosyncline."

John Fleming of the Carnegie Institution of Washington proposed that the magnetism locked into minerals could be used to understand earth movements and even to date rocks. He described a curiosity: the magnetization of certain rocks, instead of pointing toward the North Pole, pointed 180 degrees away, toward the South Pole, as though the poles had reversed.

The conference participants must have felt that their science was on the verge of a breakthrough, poised to leave behind forever the tired, descriptive past. As they departed the continents and set sail, seeking their data upon the sea, perhaps a few of the new breed of marine earth scientists thought of other great seafarers: Columbus, Magellan, Cook. How eagerly they must have anticipated the coming years of active research on and beneath the world's oceans. Indeed, at the end of the summer of 1939, the International Geological Congress was to meet in Washington, D.C., hosted by Field. Perhaps at this meeting the breakthrough provided by geophysical data would spread from the pioneering voyagers to the broad community of geologists. But that September, Hitler invaded Poland, and "many delegates turned around mid-voyage, and others . . . quickly returned home." Many joined the effort to defeat the Axis powers for the second time in little more than twenty years; some were to make discoveries that were critical to the Allied victory.

The Bottom of the Sea

With the advantage of hindsight, it is clear that although earth scientists at the end of the 1930s had to abandon their immediate research plans, the work conducted and the data collected during World War II led directly to the revolution in the earth sciences of the 1960s. It seems impossible to escape the conclusion that were it not for wartime research, the revolution would have been delayed, though by how much we cannot know.

None of the scientists who took part in the wartime and immediate postwar research in geophysics saw it as having anything to do with continental drift. Like their contemporaries, they had either rejected drift or forgotten about it. What motivated them was the thought that, until they better understood the two thirds of the earth's surface that the oceans covered, they could hardly understand the geology of the entire globe. Few could have foreseen that research in the ocean basins would quickly prove more decisive for drift than had a century of geological research on the continents.

As the historian of science John A. Stewart has pointed out, by the beginning of the 1950s, it was routine for the burgeoning number of oceanographic research vessels to use instruments developed during and just after the war to measure the following:

- Seafloor topography using sonar.
- The strength of the magnetic field using towed magnetometers.
- The pull of gravity using gravimeters greatly improved on those of Vening Meinesz.
- The structure of the oceanic crust using seismic reflection and refraction.
- The composition of the seafloor using cores recovered by drilling.
- The flow of heat through the ocean floor.

Prior to World War II, though the laying of submarine cables had revealed that some sort of ridge existed near the middle of the Atlantic, scientists thought that the rest of the ocean floor was a flat, almost featureless plain. Shortly after World War II, enough data had become available for two scientists at the Lamont Geological Observatory, Bruce Heezen and Marie Tharp, to make a detailed map of the topography of the Atlantic Ocean floor. To the shock of nearly the entire community of geologists, the Atlantic floor proved to be more rugged than the average continental surface. Down the full length of the Atlantic basin runs an enormous submarine mountain chain, undulating so as to remain equidistant from South America and Africa. Down the center of this mid-ocean ridge runs a deep rift. Hundreds of valleys, each offset horizontally, cut the ridge-rift system into segments. The deepest ocean water lies not in the expected place—somewhere near the center—but under the peripheral trenches. No one could fail to notice that by sliding South America and Africa back together as Wegener had proposed, the mid-Atlantic ridge would snuggle comfortably in between.

The map revealed a bizarre surface, one that had been there all the time, but hidden under miles of seawater. The ridge-rift system seemed to have been first horribly twisted, then riven into myriad pieces, as though sliced by the hand of a demented, cleaver-wielding giant. Something terrible had happened to the Atlantic Ocean floor.

The seafloors held more surprises. The highest heat flow escaped right over the mid-oceanic rift, while the lowest came from the trenches. The oceans had existed for at least hundreds of millions of years; during all that time, a layer of sediment several miles thick should have accumulated on the ocean floor. Yet the drill cores found only a fraction of the expected sediment thickness. Either the oceans were much younger than had been thought, or some process had swept them clean of sediment.

Almost all the shallow earthquake epicenters lay right under the mid-ocean rift. Deep-foci quakes lay far away, under the marginal trenches, clustered along a descending plane. This dipping plane might be a fault surface that defines the zone of contact between a down-welling convection cell and the crust above. Seismic studies also revealed that in the upper mantle, down about 100 kilometers or so, earthquake waves slow down, indicating that the rock at that depth is hotter. This area of low seismic velocity might be the plastic zone over which the continents move: the top of the conveyor belt.

The topography of the Atlantic floor, its heat flow, seismicity, and lack of sediment cover, were each consistent with mantle convection as Holmes had outlined it in 1929 and 1944. But a convecting mantle provided only indirect evidence of drift; could the divisive fit between the continents provide direct evidence?

Wegener, Schuchert, Willis, du Toit, and others had debated whether the fit between Africa and South America was good enough to constitute proof of drift. Some said the fit was poor; Willis said it was too good to be true; Schuchert adopted both positions in the same article. Sir Harold Jeffreys of Cambridge, writing to *Nature* in 1962, said that "Wegener's alleged fit of South America into Africa is a misfit by about 15°. This is obvious on inspection of the globe." Cambridge University legend has it that this silly assertion led Edward Bullard, a Cambridge geophysicist, to use a computer to test how well the two continents fit together. Instead of matching the continents at their coastlines, which are merely an artifact of present sea level, Bullard fit them at the break between the continental shelf and the deeper continental slope. After the computer had determined the best fit, he plotted the results on a map that highlighted the regions where the two continents overlapped and others where they fell short.

A glance at Bullard's map showed that the overlaps and gaps were small and essentially offset each other. Since in anyone's scenario, the two continents had been apart for scores of millions of years, during which erosion and deposition have modified their edges, the fit was better that even the strongest proponent of drift could have hoped. Like the map of the corrugated Atlantic floor, the "Bullard fit" made a great impression on the growing number of geologists whose minds were beginning to open to the possibility of drift.

But even the excellent fit did not establish a previous connection between the continents; to the anti-drifters, the fit was merely another piece of circumstantial evidence. More convincing corroboration would come from showing that the continents had once been located at places on the globe other than where we now find them. If the magnetism of rocks could serve as fossil compass needles, it might provide the test.

Paleomagnetism

Since about 500 B.C, people have known that a freely suspended sliver of lodestone (the mineral magnetite) points toward the North Pole. Today, we know that when an igneous magma cools, it crystallizes, and then, as it cools further, reaches a temperature at which electrons in the atoms of iron in magnetite and certain other minerals line up and lock onto the direction of the prevailing magnetic field. Thus, in theory, its magnetism might provide a permanent record of the direction to the North Pole at the time an igneous rock solidified, and so permit an objective test of drift. But first, several questions had to be answered:

- Do the magnetic minerals in rocks reliably and consistently point to the North Pole at the time the rocks form?
- Does the magnetization in rocks remain stable for hundreds of millions of years? Might not metamorphism, chemical reactions, and the like reorient the magnetic minerals?
- Half the rocks tested have magnetization that points toward the South Pole. Did the earth's magnetic field reverse its polarity, or can minerals change their own magnetic orientation spontaneously?

- We know that the position of the magnetic North Pole on the earth's surface does not coincide with that of the geographic pole, and that over time, the magnetic pole wanders randomly. Given its instability, how can we be sure that the pole has not wandered by larger deviations in the past? If the poles move, how do we tell the difference between wandering of the poles and drifting of the continents?

Patrick Blackett, a British geologist who won the Nobel Prize in Physics in 1948 for work done at Cambridge on cosmic rays, led the new field of paleomagnetism. During World War II, he and Edward Bullard had worked on magnetometers for minesweepers, and afterwards Blackett got the idea that magnetism might be a fundamental property of matter, like gravity. To test his hypothesis, he built a sensitive magnetometer and conducted experiments, but failed to find the predicted effect. Blackett accepted the negative evidence and immediately put his magnetometer to work measuring the magnetism of rocks. In order to learn the rudiments of geology, Blackett, like a generation of British geologists, read Holmes's textbook.

Blackett and Bullard each set up research groups and began experiments to test the validity and reliability of paleomagnetism. They found that rock magnetism is stable over geologic time, eliminating one problem. They also found that rocks that showed the reversed magnetism always pointed directly toward the South Pole. This allowed them to determine the position of the earth's magnetic field without regard to whether a specimen showed reversed magnetism. The short-term wandering of the magnetic pole appeared to average out to the same position as the geographic pole. These questions answered, the paleomagnetists were ready to apply their new technique to test continental drift.

The very first results, on their own British rocks, were consistent with drift: Britain appeared to have once been farther to the south but over time to have moved north and rotated 30 degrees. The trouble was, these early measurements gave only the relative position. It was equally possible that Britain had stayed in the same place while the pole shifted. Which had moved: Britain, or the pole?

The obvious way to tell was to measure the paleomagnetism of rocks from at least one other continent. If, for example, Europe and North America had remained fixed in place while the pole moved, then rocks of the same age from each continent would show the pole in the same place—albeit a different place than at present. Conversely, if the pole had remained fixed while Europe and North American had moved independently, rocks of the same age from each continent would show the pole at a different place. When the test was completed, North America and Europe each showed the pole at a different place at the same time.

Thus, the continents, and not the poles, have moved. The fit of the continents, their geological newsprint, and their magnetism all suggested the same conclusion: drift. But conclusive evidence was about to arrive. Surprisingly, it came from the strange magnetic reversals.

10

Seafloor Spreading

How extremely stupid not to have thought of that.

—Thomas Huxley

The Paleomagnetic Time Scale

French geologists, in the early 1900s, were the first to discover magnetic reversals. Subsequent research showed that a rock's magnetization was as likely to point toward one pole as the other. But the magnetization almost always pointed directly at either the North Pole or the South Pole; hardly ever in between.

As paleomagnetic research continued into the 1950s, the strange reversals increasingly appeared to be a possible fatal flaw. Had the earth's magnetic field somehow reversed itself, or had the field stayed the same while individual mineral grains spontaneously reoriented their own magnetism? Neither possibility lent paleomagnetism credence. Geophysicists could not rule out reversals of the entire field, for they could not explain even why the earth has a magnetic field, though most believed it to derive from the effect of the earth's rotation on its partly molten core. If, on the other hand, minerals could change their own magnetism, how could one be sure they would reliably continue to point to past pole positions? The possibility took on added weight when theorists identified several ways in which magnetic minerals could "self-reverse." Blackett and the other pioneers were inclined to think that the reversals were self-induced. They quickly came up with a method to differentiate between the two possibilities.

If the magnetism in rocks of the same age but from different locations always has the same polarity—either normal (like today) or reversed—then the entire field has reversed. If, on the other hand, some rocks point toward one pole while others of the same age point toward the opposite pole, then it is more likely that the minerals have reversed themselves.

The most highly magnetized rocks are volcanic; the best way to date them was to use the potassium-argon method. This technique, developed in the early 1950s at the University of California at Berkeley by Garniss Curtis and Jack Evernden, gave as one of its first results the age of the Olduvai Tuff in Tanganika, which established the age of the fossil hominid *Zinjanthropus* (later renamed *Australopithecus boisei*) at 1.75 million years. By the early 1960s, the U.S. Geological Survey had established a laboratory in Menlo Park, California, to investigate paleomagnetism and had hired three geologists to staff it. Allan Cox and Richard Doell trained as paleomagnetists at the University of California at Berkeley; G. Brent Dalrymple learned to use the potassium-argon method at the same university. Against the advice of their mentor, Professor John Verhoogen, who thought that searching for field reversals was a waste of time, the three began a program to date volcanic rocks by the potassium-argon method and to measure their magnetic directions. They were not trying to confirm or deny continental drift. Like most young American geologists of the 1950s and early 1960s, if they had reason to think of drift at all, they found the evidence unconvincing and went on to other research problems. They were interested in paleomagnetism for its own sake.

From work done at the trio's Menlo Park laboratory, together with that performed by a group at the Australian National University led by Ian McDougall, the paleomagnetic time scale emerged. It soon became clear that, with rare exceptions, rocks of the same age do have the same polarity. A chart of magnetism versus age for volcanic rocks made the correlation undeniable. Thus, the entire magnetic field, rather than individual minerals, has reversed. Furthermore, it has done so repeatedly.

One puzzling early exception was a basalt from Olduvai Gorge. Though its age fell within a time period of several hundred thousand years during which other rocks showed reversed magnetism, this basalt had normal magnetism. This might have been a rare instance of self-reversal, but experimental tests indicated otherwise. The researchers soon found other examples of short-lived "events" that fell within longer magnetic epochs. Sometimes a brief normal blip, such as at Olduvai, would show up within a long reversed epoch, and conversely. This made it clear that not only had the field reversed over and over; it had done so at random intervals.

By the fall of 1965, the paleomagnetists had worked out the magnetic polarity time scale for the last 4 million years. Yet to be confirmed, as the trio from Menlo Park reported that year at the annual meeting of the Geological Society of America, was a possible short event at 0.9 million years. By May 1966, they had corroborated the new event and described it in a paper that "was destined to become a historical marker in the earth sciences." The distinctive, barcode-like pattern of normal and reversed epochs, interspersed with the short-lived events, even allowed the new paleomagnetic time scale to be used to date rocks that could not be dated by one of the radiometric methods. Today, scientists measure the sequence of reversals in a section of rock of unknown age and match it to the paleomagnetic time scale. When a match occurs, the age of the rock has been determined.

With field reversals confirmed and the paleomagnetic time scale for the last 4 million years beginning to take shape, two unrelated lines of research were ready to be brought together to fuel a revolution, were anyone clever enough to do so. The event at 0.9 million years was to connect the two lines and clinch the case for drift.

Vine-Matthews-Morley

By the middle of the 1950s, oceanographic and geophysical surveys had produced a wealth of new data about earthquake foci, gravity, topography, heat flow, crustal structure, and magnetism. Several research vessels had towed magnetometers behind as they sailed wide stretches of the oceans, continuously recording the strength of the earth's magnetism.

The magnetic intensity at any point on the surface of the oceans is a combination of the earth's inherent magnetism, which stems from the core, and that of the rocks on the seafloor. When rocks magnetized in the same direction as the earth's magnetic field lie beneath the magnetometer, they add to the strength of the earth's field and cause a positive anomaly. Those magnetized in some other direction detract and produce a negative anomaly. Geologists expected a plot of the seaborne magnetic data to show a salt and pepper pattern, with positive and negative anomalies sprinkled randomly, but that is not what they found.

In several areas—the Northeast Pacific, for example—the plotted anomalies (with positive ones shown in black) showed up as a series of alternating stripes (Fig. 10.1).

Geologists spent many hours futilely staring at these magnetic anomaly maps, with their obvious stripes, trying to read meaning into the pattern, for something so definite in geology must be significant. For want of a better idea, most thought that the positive stripes came from sheetlike, vertical slabs of magnetized basalt that had intruded the upper seafloor, adding their magnetism to that of the field. Here was another mystery to add to geosynclines, mountains, deep-sea trenches, mid-ocean ridges and rifts, and the rest.

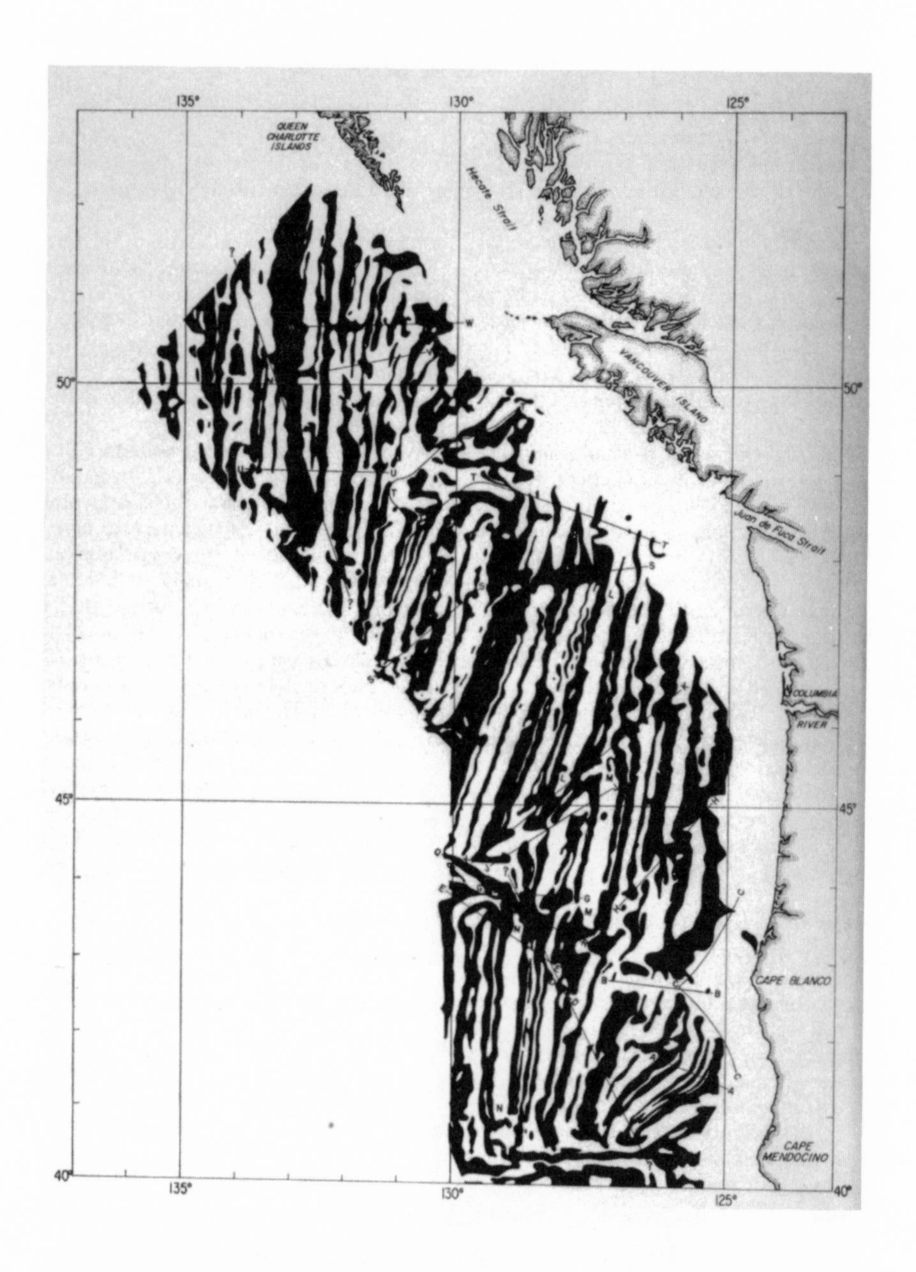

Figure 10.1 Magnetic Anomaly Patterns of the Northeast Pacific (after Raff and Mason, 1961).

In 1962, Cambridge University geophysicist Drummond Matthews was sailing the northwest Indian Ocean, recording seismic velocities and magnetization near the submarine Carlsberg Ridge. He knew that the basalts of the seafloor recorded the magnetic direction of the earth's field at the time they crystallized. Matthews returned to Cambridge with one of the best magnetic surveys of the ocean floor ever done, to find assigned to him a young graduate student named Fred Vine. Since graduate students need to be kept occupied, Matthews handed Vine the data he had accumulated for the Carlsberg Ridge, asked him to produce a map that combined the magnetic and gravity data with the submarine topography, and left on his honeymoon. "We had no idea of connecting seafloor spreading and reversals—at least I didn't," Matthews recollected.

Vine had just heard a talk given at Cambridge in January that year by Harry Hess, professor of geology at Princeton and youthful shipmate of Vening Meinesz. Hess said that the seafloor might be spreading laterally to either side of the mid-oceanic ridges. Hess's ideas so inspired the young Vine that he gave a special address to the Geology Club at Cambridge entitled "HypotHESSes."

In 1959, Hess had prepared a manuscript outlining his thinking about the seafloor and submitted it for inclusion in a book to be edited by the head of the department at Cambridge, who distributed the article to students. But the book was so long delayed that Hess eventually withdrew his paper and placed it instead in a special work published in 1962 by the Geological Society of America.

In 1961, in between Hess's original submission of his article and its publication in that book, a self-described journeyman geologist named Robert S. Dietz wrote a paper in *Nature* in which he used the term "spreading sea floor," later reversed to read "sea floor spreading," for the first time. Dietz was an American iconoclast who believed not only in seafloor spreading, but in meteorite impact as a geological force. He confronted the establishment with bizarre, non-uniformitarian claims, such as that impact created the giant Sudbury ore body in Canada and many other terrestrial craters.

In his article, Dietz wrote that "large-scale thermal convection cells operate in the mantle," welling up beneath the mid-ocean rises, spreading laterally, and turning downward at the peripheral trenches. Dietz said that the crust rides on the upper boundary of the cells, the hot, plastic zone of low seismic velocity scores of kilometers deep. Blocks of continental crust are "rafted to down-welling sites" by the convection cells, where they sit, as unsinkable as Molly Brown. Continents do not "sail like a ship" through the oceanic crust—they either "move along with it or stand still" over a down-welling site.

As to the magnetic anomalies of the Northeast Pacific, Dietz said they show "a striking north-south lineation which seems to reveal a stress pattern. Such interpretation would fit into spreading concept [sic] with the lineations being developed normal to the direction of convection creep." Reading Dietz's words with decades of hindsight, it is impossible to resist urging him onward. Just another thought or two would have made the connection that would sweep the geological community five years later: the seafloor, spreading as the field reverses, directly generates the magnetic stripes. But Dietz stops just short.

As Pasteur wrote, "chance favors only the prepared mind." Vine had read not only Hess's preprint and Dietz's article, but most of the relevant literature. He was at Cambridge with Matthews and Bullard, in Britain where opposition to continental drift was more muted than in North America. As directed, Vine set to work to interpret the magnetic patterns of the Carlsberg Ridge, beginning with two underwater seamounts that Matthews had mapped in detail. Vine was one of a small number at the time who accepted that the earth's magnetic field had reversed itself more than once. He found that he could interpret the magnetism of the seamounts if he assumed that one had normal magnetization and the other had reversed magnetization. The simplest explanation of the magnetic patterns was that one of the seamounts had formed when the magnetic field was normally oriented, as it is today, and the other when it was reversely oriented. Vine conjectured that half the oceanic crust might be reversely magnetized, and, since he also "believed in spreading," it was "a fairly small leap" to connect the magnetization of the seafloor to its spreading. Graduate student Vine thought it would look impressive if "Bullard and Vine" wrote the paper in which his speculations were presented; but Bullard, perhaps anticipating the lukewarm response, said "no way."

Vine and Matthews published "Magnetic Anomalies Over Ocean Ridges" in *Nature* on September 7, 1963. Their concept was "virtually a corollary of current ideas on ocean floor spreading and periodic reversals in the Earth's magnetic field." The idea was so simple that they could summarize it in two sentences:

> If the main crustal layer of the oceanic crust is formed over a convective up current in the mantle at the center of an oceanic ridge, it will be magnetized in the current direction of the Earth's field. Thus, if spreading of the ocean floor occurs, blocks of alternately normal and reversely magnetized material would drift away from the center of the ridge and parallel to the crest of it.

In this hypothesis, the magnetic anomaly stripes represent alternating belts of normally and reversely magnetized volcanic rock that froze on the ridge, were magnetized, and then were pulled to either side. Normally magnetized rocks add to the strength of the earth's field and cause a positive anomaly; reversely magnetized rocks subtract and cause a negative anomaly.

Vine and Matthews tied together two independent and previously unrelated concepts: seafloor spreading and magnetic field reversals. In 1963, hardly anyone believed simultaneously in both. Most continued to think that the reversals were likely to be self-induced and that the geophysical evidence contradicted seafloor spreading. As a result, few paid attention to the Vine-Matthews paper. One year after publication, Vine thought the hypothesis was going over "like a lead balloon." He was "getting pretty discouraged and beginning to lose faith."

He had good reason. At a Symposium on Continental Drift held in London in 1964, with most of the leading lights present, only one person even mentioned the Vine-Matthews hypothesis, and then only to say that it was "probably not adequate to account for all the facts of observation." John Verhoogen, mentor to the paleomagnetists, said later that at the time he had "thought Vine and Matthews' hypothesis rather ridiculous." Even as late as November 1966, Maurice Ewing, an American marine geophysicist and director of the Lamont Geological Observatory of Columbia University, drew Bullard aside at a meeting to say, "You don't believe all this rubbish, do you, Teddy?"

It often happens in science that two individuals or groups working apart and unbeknownst to each other come to the same conclusion simultaneously. In 1963, Lawrence Morley was chief of the Geophysics Division of the Canadian Geological Survey, in charge of magnetic surveys of the continental shelf off Newfoundland; thus, he understood paleomagnetism. Morley said later that he had in front of him both the paper showing the magnetic lineations in the Northeast Pacific and Dietz's seafloor spreading paper, making it "a natural thing that the two should go together."

Morley wrote up his idea—that the seafloor, spreading as the magnetic field reverses, directly generates the magnetic stripes—and submitted it as a letter to *Nature* in February 1963, months before Vine and Matthews submitted their article to the same journal. The editors replied that they "did not have room to print" the letter; in April 1963, Morley submitted it to the *Journal of Geophysical Research.* Finally, in late September 1963, a few weeks after the Vine-Matthews paper had appeared, Morley heard from the editor of *JGR:* "Such speculation makes interesting talk at cocktail parties, but it is not the sort of thing that ought to be published under serious scientific aegis." No major journal ever published Morley's paper, leaving credit to Vine and Matthews for the key connection between reversals and the spreading seafloor. One geologist later said that Morley's was "probably the most significant paper in the earth sciences ever to be denied publication."

The question of who deserves priority for connecting seafloor spreading and magnetic reversals is complicated, because the Vine and Matthews paper, unlike Morley's, was reviewed by peers and contained data. If Morley had been clever and well connected, he would have sent his paper around as an unreviewed preprint, allowing him to assert priority even though he published later. But Morley did not do so. To be fair, we ought to refer to the "Vine-Matthews-Morley" hypothesis, but few have. In the fast-moving early days of a revolution, fairness is elusive.

Confirmation

In 1965, a vital assemblage of geologists gathered at Cambridge University, and the intellectual sparks flew. The group included the usual Cambridge cast of Vine, Bullard, and Matthews, leavened by Hess and Canadian Tuzo Wilson, both of whom were there on sabbatical leave. All by this time were anxious to close the case for seafloor spreading; jointly, they came close. Vine and Wilson collaborated in using the latest paleomagnetic time scale and a computer program to see if they could simulate the magnetic anomaly patterns in the Northeast Pacific. They substituted different rates of spreading from the ridge axis into their program (they had no way of knowing the actual rate) and found that an assumed rate of 2 centimeters per year brought the simulated magnetic profile into close, but not perfect, correspondence with the actual profile. They could achieve a better match by fiddling with different spreading rates at different times, but that appeared contrived. Something was not quite right.

Vine and Wilson published their results in October 1965, a month before the annual meeting of the Geological Society of America (GSA). Meanwhile, the Menlo Park team of Cox, Doell, and Dalrymple had been hard at work dating and measuring the magnetic polarity of volcanic rocks of different ages. By the mid-1960s, they knew that the field had reversed repeatedly; now they had to establish just when it had done so. The discovery of the short magnetic "events," so brief as to be easily missed, complicated their task.

To be complete, the paleomagnetists had to extend their time scale as far back as the resolution of potassium-argon dating would allow, and it had to include all the reversals. In order to avoid missing one or more of the short-lived events, the team had to find and date volcanic rocks that covered the complete range of ages back at least 4 million years. By 1964–65, they had analyzed fifteen specimens in the range between 1.0 and 1.5 million years, but only one specimen between 0.7 and 1.0, and that one happened to have reversed magnetism. To fill this critical time gap, Doell and Dalrymple went searching for rocks of the right age.

They found them in a suite of volcanic rocks from the Valles Caldera in the Jemez Mountains of New Mexico, north of Santa Fe. They selected six samples from up and down the volcanic suite. Three dated to near 0.7 million years and were reversed; one dated to 0.88 million years and had an intermediate polarity, as though it had frozen while the field was reversing; one dated to 0.89 million years and had normal polarity; the oldest, at 1.04 million years, had reversed polarity. Thus, during the middle of a long reversed epoch, the field had flipped briefly to normal and then flipped back again.

At the GSA meeting in November 1965, Brent Dalrymple presented the results of the latest paleomagnetic research from his group, noting that "There may be another event at about 0.9 million years, although it is not yet confirmed." They were to name the event the Jaramillo, after a small creek near the site of the critical sample. Fred

Vine was present at the meeting, and recalled that when he heard of the Jaramillo event, he "realized immediately that the [Northeast Pacific anomaly patterns] could be interpreted in terms of a constant spreading rate"—there was no need to fiddle with different spreading rates at different times. Not knowing of the Jaramillo event, he and Wilson had been trying to force a match between the simulated and the actual magnetic patterns, but with a critical magnetic reversal missing. With the Jaramillo event added, the real and simulated profiles came into perfect concordance.

Recognition of the Jaramillo event not only brought the calculated and actual profiles in the Northeast Pacific into identity; it revealed a correspondence that no scientist could ignore: the paleomagnetic time scale, when combined with the assumption of a constant rate of spreading outward from a ridge, precisely predicted the pattern of seafloor magnetic anomalies around mid-ocean ridges. Scientists working apart from each other, barely aware of each other's existence, had developed the paleomagnetic time scale, the theory of spreading, and the seafloor magnetic anomalies. For those who had denied drift and seafloor spreading, the horns of a dilemma became uncomfortably sharp: either they had to change their mind publicly or ascribe the apparent corroboration of the Vine-Matthews-Morley hypothesis to coincidence.

Scientists at the Lamont Geological Observatory of Columbia University collected most of the data that would eventually confirm seafloor spreading and plate tectonics. Yet the director of the Observatory, Maurice Ewing, remained a staunch anti-drifter. Few Lamont scientists dared cross him. One graduate student recalled, "Ewing's philosophy that the ocean floors were permanent features was the party line at Lamont; seafloor spreading was anathema!" Euripides told us that "The gods visit the sins of the fathers upon the children." Columbia University professor Walter Bucher, a prominent structural geologist and adamant anti-drifter, had strongly influenced Ewing, who passed on his own opposition to drift to his Lamont progeny. According to Robert Dietz, "Ewing remained a fixist until the bitter end at the 1967 American Geophysical Union meeting." But the Lamont scientists continued to collect their geophysical data. Ultimately, in spite of magisters and dogma, data do not lie.

A research cruise to the deep South Pacific by the National Science Foundation vessel *Eltanin* provided conclusive evidence for seafloor spreading and set the stage for plate tectonics. On Leg 19, the *Eltanin* crossed the East Pacific Rise, the broader, lower analog of the mid-Atlantic ridge, at 50 degrees South latitude. Although the time was late 1965, none of the scientists aboard were aware of the Vine-Matthews hypothesis. They were at sea to make seismic measurements, but decided as an afterthought to gather magnetic data as well. The scientist responsible for plotting the magnetic intensities as a profile across the ridge axis was an American graduate student named Walter Pitman. He had shown another magnetic profile from the *Eltanin* cruise to his Lamont colleagues, noting that it looked "almost exactly the same as the profile on the Northeast Pacific ridge," which had come out a few months before. At least one of these colleagues had known of the Vine-Matthews hypothesis, for he said, "Ha! Ha! I suppose you're going to prove Vine and Matthews are right." There followed further ridicule of Dietz, Hess, and seafloor spreading in general. As Pitman continued his data analysis, he saw that the magnetic profile that emerged from *Eltanin* Leg 19 was unusually symmetrical: the highs and lows of magnetic intensity outward from the center of the ridge on one side exactly matched those on the other.

James Heirtzler led the *Eltanin* project. In 1965, he concluded that most magnetic profiles "do not follow the pattern assumed by Vine and Matthews." When he saw the Leg 19 profile, Heirtzler said it was "too perfect" to fit the Vine-Matthews hypothesis. Another Lamont scientist, after examining the profile, said, "Well, that knocks seafloor spreading into a cocked hat. It's too perfect." The reaction of one of the USGS trio of paleomagnetists, Richard Doell, was different. "It's so good it can't possibly be true," Doell commented. Then he added, "but it is."

In February 1966, Fred Vine arrived at the data-filled labs of Lamont, in William Glen's felicitous phrase, "a starving man let loose in the kitchen of the Cordon Bleu." Unconvinced until that moment of his own hypothesis, when Vine saw the *Eltanin* Leg 19 profile, he knew that "it was all over but the shouting."

In a spirit that exemplifies openness in science, Heirtzler later provided the *Eltanin* magnetic data to Vine, who was then working with Hess at Princeton, so that Vine could study the profile in detail. By the time Vine returned to Lamont a few months later to give a talk, it was obvious that he was preparing a major paper based on the *Eltanin* data. Pitman suggested that they publish jointly, but Vine said no. As Pitman put it, "Well, God, *he* knew what it meant—why should he share that with anybody?" Finally, all parties agreed that the Lamont scientists would publish a shorter article two weeks before Vine published a more complete one.

Most who have studied the plate tectonic revolution mark the April 1966 meeting of the American Geophysical Union (AGU) in Washington, D.C., as the true beginning. Heirtzler presented the *Eltanin* profile at a conference session with paleomagnetist Allan Cox in the chair. Cox recognized in the seafloor profile not only the Jaramillo event, but each epoch, event, and boundary that his group had found recorded in volcanic rocks on land. The conveyor belt of seafloor spreading turned out to have been a magnetic tape recorder. Cox later described 1966 as "the most exciting year of my life."

Vine came to the 1966 AGU meeting armed with a preprint of the article he was preparing to submit to *Science.* Using the latest paleomagnetic time scale and a different but constant spreading rate for each ridge, Vine generated the observed magnetic profile across the mid-Atlantic, Reykjanes, Red Sea, Carlsberg, Juan de Fuca (NE Pacific), Gorda (NE Pacific), and East Pacific ridges. When the paper appeared (in *Science* on December 16, 1966), Vine concluded that "The entire history of the ocean basins is contained frozen in the ocean crust. The rates [of spreading] obtained are in exact agreement with those needed to account for continental drift."

Further confirmation of the Vine-Matthews-Morley hypothesis came from an unexpected source: the deep-sea drill cores that for years Lamont scientists had been collecting and storing. As the techniques of paleomagnetism had improved, it became possible to determine the direction of the magnetism in sediments and sedimentary rocks, even though it was thousands of times weaker than the magnetism of volcanic rocks. Neil Opdyke, one of the few Lamonters who maintained an open interest in continental drift, used an ultra-sensitive magnetometer to measure the paleomagnetism in cores in the Lamont collections. He focused on a set from the deep South Atlantic, paleomagnetic directions being easier to determine at high latitudes. As the work progressed, Opdyke detected in the cores each reversal of the paleomagnetic time scale. In the autumn of 1965, he located a new, short-lived reversal at 0.9 million years, but was unsure whether it was real. By the pivotal spring 1966 AGU meeting, he was more confident. In a hotel room discussion, Opdyke told Dalrymple of the new event, only to hear him respond, "Too late—I can tell you the name. We've called it the Jaramillo event."

Undaunted, and with access to the best deep-sea cores, Opdyke returned to Lamont and began months of "fever-pitched activity." He and his colleagues reported their results in *Science* on October 21, 1966, two months before Vine's paper. In each of eight deep South Atlantic cores, they detected each of the major magnetic reversal boundaries and each of the short events. The ages of different sections of the cores as deduced from paleomagnetism matched those determined from microfossils. Everything dovetailed.

The deep-sea cores provided a third base of support for seafloor spreading. Lava flows extruded from volcanoes on land, basalts erupted onto the seafloor, and sediments that had settled through miles of ocean water to accumulate on the bottom of the sea—all recorded the same magnetic field reversals. As Tuzo Wilson expressed it, "Three different features of the earth all change in exactly the same ratios. These ratios are the same in all parts of the world. The results from one set are thus being used to make precise numerical predictions about all the sets in all parts of the world." Here was either cause and effect, or coincidence of galactic proportions.

Between 1966 and 1968, scientists from many disciplines rushed to reinterpret their data in light of seafloor spreading. Several enunciated the idea that the crust of the earth comprises a set of rigid, interlocked plates, each moving as a unit. Two plates can converge, split apart and separate, or slide past one another. Earthquakes arise mainly at the boundaries where plates rub against each other.

Lamont scientists now embraced the previously anathematic rubbish of seafloor spreading and began a furious effort to exploit their vast databank. In an article in *Scientific American* in 1968, Heirtzler showed that the magnetic stripes continue out from the ridge crests, far beyond the 4-million-year limit of potassium-argon accuracy. The North Atlantic, the South Atlantic, and the Pacific Antarctic Oceans all have the same pattern of anomalies stretching back for nearly 80 million years, each ridge with its own, constant spreading rate. The match was so good that the process of dating anomalies could be turned inside out: now the magnetic reversal time scale back to 80 million years could be deduced from the seafloor anomalies.

A final piece of evidence in their own language convinced almost all continental geologists. During the 1960s, research vessels recovered cores that reached all the way to the bottom of the sedimentary layer that lies above the hard basalt "basement" of the seafloor. Paleontologists dated the lowest, oldest sedimentary rocks in the cores using fossils, and found that the farther away the bottom sedimentary layer is from the ridge, the older it is. The age of the seafloor at any point, as determined from the deepest part of the cores, turned out to correspond exactly to that derived from the magnetic anomalies. The spreading rate calculated from the sediment ages and the one calculated from paleomagnetism were the same.

Every geologist could understand the deep-sea drilling results. Samples were collected from a known point (albeit on the seafloor), brought into a laboratory, and dated using the time-tested methods of paleontology. The results produced a geologic map of the seafloor. Nowhere was the floor of the Atlantic Ocean older than about 180 million years.

The next step was to transfer plate tectonics from the floor of the sea to the surface of the continents. In his 1966 paper in *Science,* Vine began the process by showing how the westward drifting North American continent had overridden the former trench system of the East Pacific Rise. His analysis placed the crest of the rise right under the Colorado Plateau, which within the last few million years has undergone a rapid and mysterious uplift.

In the mid-1960s, Robert Dietz showed how to wed the concept of geosynclines to the new tectonics, but as usual was ahead of his time. In 1970, John Dewey and F. M. Bird showed how moving plates can explain every major feature of mountain ranges. Geosynclines are not gradually sinking troughs, but are wedges of sediment that accumulate on the trailing margins of receding continents. When the two adjoining plates reverse direction, converge, and collide, their facing sediment wedges crumple into a mountain range.

In 1968, Tuzo Wilson suggested that the margins of continents might contain pieces broken off from older continents and carried along until the conveyor belt turned downward and plastered the fragment onto a new continent. Today, geologists think that as much as one fourth of North America may be composed of such welded microcontinents, thus turning the jigsaw puzzle into a patchwork quilt. John McPhee has captured well how geologists of the 1970s and 1980s regarded these "exotic terranes."

Peering back through the most recent assemblage of continents, geologists see glimmers of ancient oceans that existed before the Atlantic. Ocean basins, far from being permanent, as Dana and Willis had pronounced, are ephemeral. As continents converge, ocean basins shrink and disappear; when continents rift apart and seafloors spread, new ocean basins are born and grow, eventually to die.

Since some plates carry continents and others do not, three types of collisions between plates take place: ocean plate to ocean plate (as in Japan, the Aleutians, and the Philippines); ocean to continent (Andes, Cascades, Sierra Nevada); and continent to continent (Alps, Urals, Himalayas, Appalachians). Each type produces a different set of features, more than enough to explain the broad tectonics of the earth's surface.

In the thirty years between the advent of plate tectonics and the end of the twentieth century, the theory has unified information from many subdisciplines. Not only does plate tectonics elucidate the history of the ocean basins; it explains the evolution of continents, the nature and fate of geosynclines, the formation of mountains, the origin of volcanic rocks, and even the reason for the existence of the continental crust.

Plate tectonics explained nearly everything except what moves the plates. The scientists of the 1960s accepted mantle convection as the cause of plate motion, but none knew exactly how it worked. Since the plates lock together such that if one moves, all the others have to move as well, it is difficult to work out what is happening in the mantle from the movement of plates on the surface,.

Convection might drive the plates in one of several ways, or in some combination. One possibility goes back to Reginald Daly, who in 1923 postulated a mechanism that he called "gravity sliding" and today we call "slab-pull." As the cold, downturning plate descends at the margins of the ocean basins, it drags the entire crust on one side of the ridge along with it and pulls the faraway ridge crest apart, allowing new magma to enter the expanding mid-oceanic crack and spread out to either side. Another possible mechanism is that magma rising beneath a ridge pushes the crust to either side to make room for itself, thus forcing the plates apart: ridge-push. A third is that "sticky" convection cells adhere to the bottom of the plates and tow them along: plate drag. No one knows exactly how convection moves the plates. But since plates do move, they can.

Dissent

Such a large amount of new, hard data—all mutually supportive—backed up plate tectonics that from the outset, dissent was rare but enthusiastic. What the magisters of geology thought of the Bullard fit, the apparent polar wandering curves, the *Eltanin* Leg 19 profile, Opdyke's sediment core reversals, the age of the seafloor, and so on seemed no longer to matter. The data were plain for anyone, even the rawest student, to see. Geology had long since left behind science by rhetoric; now it became democratized. But, of course, not all the magisters agreed.

Sir Harold Jeffreys, in the 1970 edition of his venerable book *The Earth,* took the same position he had taken in the first edition forty-six years before: the laws of physics render drift (and along with it, convection, seafloor spreading, and plate tectonics) impossible. Russian geologists, working in isolation from the West and far from the continental margins where plates come together to evince their existence, were also opposed. Hearkening back to Suess, they had developed an elaborate theory of vertical tectonics, in which blocks move, not sideways, but up and down along huge faults, the Colorado Plateau being a good example. Their champion was a distinguished and internationally known Russian academician, Vladimir Vladimirovich Belousov. To his credit, in 1968 Belousov engaged Tuzo Wilson in a gentlemanly debate over plate tectonics.

In the early 1970s, a father-and-son American team, A. A. and Howard A. Meyerhoff, delivered a series of articles in which they attempted to show that plate tectonics did not explain as much as its proponents claimed, was full of contradictions, and included only data that supported it. They argued that the deep-sea drilling project might not have reached basement level and therefore might not have extracted the oldest sediments from the seafloor. They pointed to a long-recognized anomaly: a slice of a 100-million-year-old section of ocean floor near the supposedly young mid-Atlantic ridge. As an alternative to plate tectonics, they resurrected the old theory of contraction.

A. A. Meyerhoff also joined the losing side in a geologic debate of the 1980s and 1990s: whether a meteorite struck the earth and ended the long reign of the dinosaurs. He and his coauthor, Charles Officer of Dartmouth, derided the claim that the so-called K-T crater had been located in the Yucatan, labeling the idea a "Caribbean Caper." Peering behind the veil of their derisive wording, one again detects the long shadow that uniformitarianism has cast over twentieth-century geology.

Thomas Kuhn's theory of scientific revolutions holds that, as new evidence emerges, a scientific paradigm requires a succession of adjustments. As the modifications accumulate, the sum of the tinkering sometimes produces a Rube Goldberg–like contraption whose failure seems imminent to all but the most dogmatic and faithful. At about this point, a rival theory emerges. Eventually the rival, or some other theory, replaces the old, though not in the minds of the older theory's more senior supporters, who, as Max Planck predicted, usually must die first. Only one who ignored history could fail to ask: What are the odds that a better theory will someday replace plate tectonics?

In answering, one must distinguish fact from theory. We discover facts and invent theories to explain them. But the line between fact and theory is not always clear. Over time, as evidence accumulates, a corroborated theory can begin to take on the aspect of fact. Is plate tectonics still a theory, or is it now a fact?

The fair answer seems to be that, regardless of the driving mechanism, by the early years of the twenty-first century, continental drift and plate tectonics have become observational facts. Using modern instruments, geologists can measure the previously imperceptible movement of the continents and plates, thus realizing Wegener's old hope. A vast amount of interlocking data confirm plate tectonics. To geologists today, that plates move and carry continents with them seems as much a certainty as that tomorrow the sun will rise in the East. Admittedly, while we know why the sun appears to rise where it does, we still do not know exactly why plates move.

11

Rejection and Priority

"Holmes was an anachronism in our time, because he knew so much."

—PRINCETON GRADUATE STUDENT

Hidden Doubts

In their scientific articles and books, the geologists who wrote about continental drift appeared to have firm convictions. Charles Schuchert obstinately opposed drift; Arthur Holmes endorsed it. But in their private writings, or on the rare occasion when they allowed their feelings to creep into their professional articles, the two exposed the tension, even the inner turmoil, that came from wrestling with the great theory. Peering into their rarely revealed inner thoughts is like entering the mind of a person who has begun to doubt a long-held religious faith.

F. Scott Fitzgerald unknowingly endorsed Chamberlin's principle of multiple working hypotheses when he said, "The test of a first-rate intelligence is the ability to hold two opposed ideas in mind at the same time and still retain the ability to function." But to preach such lofty ideals is easier than to practice them. At the very moment when

Schuchert was writing his influential, anti-drift paper for the AAPG Symposium, doubt filled his mind.

In a letter to Holmes in 1926, Schuchert wrote that he could not see why "Brazil cracked away from Africa . . . and . . . flowed several thousand miles away to the west." But if it had, Schuchert wrote, "We can explain all the more easily the Permian glaciation and the [South American] faunas." In 1929, he wrote, "In spite of all the . . . American criticism about Gondwana [continental drift], and the fact that I *want* to give up the theory, so far I have not been able to explain away what I know of the marine faunas." He wrote to another colleague that "It is along the lines laid down by Holmes that geologists will eventually solve the causes of deformative mountains and how they are made." On the same day, he wrote to another, "I completely reject the wholesale sliding theory of Wegener. No paleontologist can accept continental movements of the extent of thousands of miles." These mutually exclusive conclusions, held simultaneously in Schuchert's mind, expose the turmoil he must have felt. But only his private correspondents knew of his doubt; to the rest of the geological world, he appeared adamantly opposed to drift.

But what of Arthur Holmes? Surely this man, who had adopted drift in the 1920s and staunchly defended it, entertained no doubts. But in time, perhaps worn down by his lonely fight, Holmes did begin to doubt. In 1953, he reviewed the published proceedings of a 1949 conference of the Society for the Study of Evolution. Ernst Mayr, Curator of Birds at the American Museum of Natural History, chaired the meeting, and many leading paleontologists participated. Little new evidence to bear on drift had come out since du Toit's *Our Wandering Continents.* Like their predecessors at the AAPG Symposium two decades earlier, with near unanimity the paleontologists rejected drift. Walter Bucher of Columbia, mentor to Ewing, banished it entirely: "The concept of continental drift cannot be used as a working hypothesis."

After concluding his balanced review, in an unexpected final section, Holmes wrote:

> To end on a personal note: I should confess that, despite appearances to the contrary, I have never succeeded in freeing myself from a nagging prejudice against continental drift; in my geological bones, so to speak, I feel the hypothesis is a fantastic one. But this is not science, and in reaction I have been deliberately careful not to ignore the very formidable body of evidence that has seemed to make continental drift an inescapable inference.

Fighting a battle nearly single-handedly for twenty-five years may have taken a toll on Holmes's self-confidence. He even went so far as to think the previously unthinkable: "This symposium has left me with the general impression that a few land bridges or linkages by island stepping-stones would probably suffice for the biogeographical problems."

But by the end of the review, Holmes had regained his composure:

> While so many contradictory voices confuse judgment, one cannot do better than commend Dunbar's wise dictum that "it is unsafe to reject, a priori, either continental drift or foundering of broad land bridges."

With the advantage of half a century of hindsight, it is astonishing to witness that in 1949, leading paleontologists and geologists, at least in North America, continued to endorse land bridges, for which not one shred of direct evidence ever existed and which were entirely the product of their imaginations. Even Holmes wavered. A kind of groupthink had set in.

Holmes's next opportunity to record his personal views came in

1956, when he received the Penrose Medal of the Geological Society of America. In his citation of the long list of Holmes's accomplishments, stretching all the way back to the book on the age of the earth in 1913, presenter Hollis Hedberg said not one word about Holmes's career-long interest in convection and continental drift. The closest Hedberg came, near the end of the citation, was to say, "You see, I have not had time even to mention his contributions to sedimentary-rock petrology as well as to ore deposits, tectonics, and many other branches of geology." Buried in "tectonics . . . and many other branches" were the seminal contributions for which Holmes will be remembered and which today would be at the top of the list of his accomplishments.

Was Hedberg's omission of what today would seem to be Holmes's most important contributions a conscious insult? No, because Holmes apparently had a chance to put convection and drift on Hedberg's list but elected not to do so. Before writing the citation, Hedberg had done Holmes the courtesy of visiting him in Britain. He asked Holmes what research had given him the greatest satisfaction. Holmes cited two accomplishments: (1) the discovery that the radioactive isotope of potassium was number 41, which opened up a new dating technique. (When the radioactive isotope turned out to be potassium 40, this turned into Holmes's biggest disappointment.); and (2) His 1946 estimate of the age of the earth. In his gracious written response, Holmes thanked Hedberg for visiting him the previous August. It seems inescapable that, tacitly at least, Holmes approved Hedberg's citation. For whatever reason, Holmes must have endorsed the omission of continental drift and convection from his list of achievements.

Arthur Holmes's publications throughout the 1950s and early 1960s dealt not with drift, but with the origin of rocks, particularly the exotic ones from Africa that had long held his interest. Given his lengthy tenure among British geologists and his breadth, Holmes was often called on to write obituaries. His last opportunity to address the old question of drift came with the publication of the 1965 edition of his great textbook. The preface is dated October 1964, presumably the last time at which Holmes could incorporate new information. Recall that in the fall of 1964, Fred Vine had begun to doubt his seafloor spreading hypothesis and the best informed geologists rejected drift and the evidence of their own data. Given the task of revising a lengthy masterpiece, it is not surprising that Holmes did not cite most of the newly arrived geophysical data. He did mention enough evidence from heat flow, earthquakes, and polar wandering to show that they lent new credence to convection and drift. Arthur Holmes's last statement on continental drift was this:

> The Russian geologist Beloussov [*sic*] refers in scathing terms to "the total vacuousness and sterility of the hypothesis" and professes "profound amazement" that it should even have been seriously discussed. Equating the great concept of continental drift with Wegener's "hypothesis" is an obvious source of unnecessary confusion. There never was a single hypothesis in this vast field. Secondly, "sterility" is about the least appropriate and most unjust epithet that could be applied to the continental drift concept. The latter . . . has won the distinction of being among the most stimulating incentives to geological and geophysical research.

In 1964, as he was completing his vast manuscript, Arthur Holmes was awarded the Vetlesen Prize, the closest thing geology has to a Nobel, "for scientific achievement in a clear understanding of the earth; its history and its relation to the universe." He died on September 20, 1965, in the year in which the final edition of his textbook appeared. His biographer for the Royal Society redressed Hedberg's omission, describing how Holmes had "thrown in his lot" with the drifters and "encountered severe opposition . . . but in the last decade has had the satisfaction of seeing the hypothesis revived." He said that Holmes's book "deserves to rank with Lyell's *Principles* as one of the most successful textbooks of geology ever written."

Why Was Drift Rejected?

Geologists of the first half of the twentieth century frequently paid homage to the ideal of multiple working hypotheses. With the advantage of hindsight, it is clear that they had ample cause to retain drift as one such hypothesis. They had the geological newsprint evidence of du Toit. They had a mechanism: Holmes's convection. They had no rival theory to account for the earth's behavior. Yet the leading American geologists rejected drift in favor of land bridges, which at best explained only the paleontological evidence. Why?

Wegener's methods drew much of the early criticism, especially at the 1926 AAPG Symposium. Here was an outsider accidentally discovering a theory that, if true, would outrank any ever conceived by practicing geologists, and doing so while bedridden and without having done original research. Certainly, Wegener's methods counted against him at the start. And yet we ought to remind ourselves that those who rejected him were just as intelligent as we, possibly more so, and we have the advantage of hindsight. Certainly, it must have occurred to Wegener's critics that even though he used flawed methods, Wegener might have stumbled across the right answer. But neither the flaws in Wegener's pedigree nor in his methods caused Bailey Willis to denounce drift as a fairy tale thirty-two years after the theory first appeared. Wegener was long dead. The rejection of drift was no longer personal; it was deeply professional.

Two poignant statements capture the true reason for the rejection of drift. One is Rollin Chamberlin's cri de coeur at the AAPG Symposium: "If we are to believe Wegener's hypothesis we must forget everything which has been learned in the last 70 years and start all over again." The other is contained in a letter that Schuchert wrote to Holmes in 1931:

> As a paleontologist, I have no objection to your moving about the continents about a few to several hundred miles, but when you demand thousands of miles you drifters say to us workers on marine faunas that all that is known about faunal distribution is worthless. This means that the present distribution of marine faunas cannot be trusted, and cannot be applied to the life of the past.

If we excise a few of the words in the last sentence, Schuchert says that to accept drift is to admit that "the present . . . cannot be trusted, and cannot be applied to the . . . past." Yet the motto of geologists is that "the present is the key to the past": the code words for uniformitarianism. If "the present cannot be applied to the past," geologists would have no platform from which to conduct their work. In Schuchert's philosophy, to accept continental drift was to abandon uniformitarianism, and that he and his contemporaries could not do. Without uniformitarianism, they would have had no guiding principle and would have had to face the possibility that geology was not truly a science.

That drift violated uniformitarianism was reason enough for Schuchert. But as the years passed, professors of geology, when required to explain to their students the reasons for the rejection of drift, needed something more. To say that geologists denied drift because it violated uniformitarianism would have made professors appear doctrinaire and unwilling to follow their own professed principle of multiple working hypotheses. To attribute the rejection to Wegener's methodology would have made the profession appear unscientific and *ad hominem.* The reasons presented ex post facto had to be believable on their face and not damage the reputations of the rejectors or the profession.

Two explanations came to be accepted: First, Wegener had not offered a plausible mechanism. This misleading criticism we find all the way into the 1960s. Second, by appearing to explain the fossil evidence, which required that the continents once had been connected, land bridges rendered drift "unnecessary." The same response came in the 1980s to the Alvarez theory of meteorite impact, which some said was unnecessary to explain the extinction of the dinosaurs, since uniformitarian processes already explained it. Magisters, having earned their reputations within the current paradigm, naturally are inclined to view a new theory as unnecessary. But the history of science shows that new theories not only are necessary, they are inevitable.

It is hard for a modern geologist to accept that our predecessors, the giants to whom we owe so much, behaved unscientifically. But they did. Caught in the mores of their discipline, entangled by alliances personal and professional, befogged by an allegiance to uniformitarianism, understandably unwilling to "forget all they had learned," they lost their way. Like all children, we geologists have had to come to accept that our intellectual forebears were, like us, fallible human beings.

Priority

Almost four decades after the 1963 paper by Vine and Matthews, though the theory has required adjustment, and though the exact cause of plate movement remains unknown, plate tectonics is robust. That a set of large, thin, moving slabs comprise the earth's exterior is an observational fact. The French geochemist, author, and former minister of education Claude Allègre, in 1988, summed up the effect of the plate tectonic revolution: "In the midst of this intellectual banquet, geology was to accomplish more in ten years than it had in the previous one hundred years." Given the status of plate tectonics as the most important theory in geology, for the sake of the history of science and as a matter of simple fairness, it is essential to know who deserves priority for it.

From the plate tectonics revolution onward, neither Alfred Wegener, Arthur Holmes, nor Robert Dietz has received credit for the connected theories of continental drift, seafloor spreading, and plate tectonics. In science, traditionally and logically, the person who first publishes an idea, usually as an article in a peer-reviewed scientific journal, but sometimes in a book, receives credit. Casual conversation, oral presentations at scientific meetings that are not subsequently recorded and published, and preprints of articles distributed to one's colleagues, do not establish priority, and for good reason. Oral conversation leaves no record other than faulty and self-serving memory. Preprints and talks at professional meetings have not received peer review, and when they do, may be revised or rejected. Yet in the case of credit for seafloor spreading, geologists broke the rules to give priority to Harry Hess based on a preprint. Does he deserve it?

In his 1961 paper on seafloor spreading, Dietz wrote:

> The median rises mark the up-welling sites or divergences [of convection cells]; the trenches are associated with convergences or down-welling sites . .
> The spreading concept envisages limited continental drifting, with the sial [continental crustal] blocks initially being rafted to down-welling sites and then being stabilized in a balanced field of opposing drag forces.

In April 1961, Hess began to circulate to his colleagues a preprint of a thirty-eight-page manuscript entitled "The Evolution of Ocean Basins." It was to be included in a forthcoming book called *The Sea, Ideas and Observations.* Dietz's seafloor spreading paper appeared in *Nature* on June 3, 1961. No submission date is listed, but given the typical time for review and editing—at least a few months—it is likely that Dietz submitted his paper before Hess began to distribute his preprint. In any case, Dietz did not cite Hess's preprint. Indeed, he had not read it, as his longtime colleague, the oceanographer William Menard, makes clear:

> The marine geologists around San Diego in early 1961 began to weary of editing and reviewing Dietz's manuscripts. Dietz,

> however, never wrote a dull sentence in his life, and perhaps with resignation, I began to read yet another, mercifully short manuscript sometime in the spring [of 1961]. I was dumbfounded. Hess had sent me a copy of "The Evolution of Ocean Basins" soon after he bound the manuscript. Dietz's manuscript, "Continent and Ocean Basin Evolution by Spreading of the Sea Floor," was amazingly similar in more than the name. I remember what then happened 23 years ago with perfect clarity. Dietz remembers it exactly the same way. I phoned him with the news. He expressed surprise because he had not received or seen Hess's manuscript. He came over to my office to read the manuscript for the first time.

In a look back at his long, multifarious career, Dietz remembers it the same way:

> In late 1960, I initially wrote a brief speculative paper . . . which eventually appeared in *Nature* in June 1961. It was a potboiler. Unbeknownst to me, Harry Hess, independently and earlier (1960 preprint published in 1962) had suggested almost the same idea. Bill Menard received a copy of the preprint in May 1961 and contacted me since he had earlier reviewed my manuscript already in press. I agreed that Harry Hess should be accorded priority and did so in a 1962 publication.

Even if he had read Hess's preprint, since Dietz submitted and published first, scientific practice would demand that he receive credit for seafloor spreading. The first in print in a refereed journal has priority. Surely then, between Dietz and Hess, Dietz has it. This would not only be following scientific practice, it would be just: if we accept the word of Dietz, confirmed by Menard, not only had Dietz not read Hess's preprint, he had not known of Hess's ideas. But Hess did not see it that way.

For most of the twentieth century up to the early 1960s, geologists worked largely on local problems. A delay of a year or two in publishing mattered little: no one else was studying your research subject; one could take one's time and run little risk of being scooped. But by mid-1961, Dietz's paper had appeared, yet *The Sea* was dormant. For someone as tuned in as Hess, it was apparent that, finally, things were beginning to move in geology, and that if he did not hurry, his field would leave him behind and give credit to someone else. Hess wrote to the editor (who had been Hess's student) of a book to be published by the Geological Society of America, asking if the manuscript that he had submitted to *The Sea* and distributed as a preprint could be included instead in the new work, for it "is timely now but may be very dead by 1964." The editor agreed, and the paper, retitled "History of Ocean Basins," appeared in the GSA publication in November 1962.

Hess began by describing his paper as "an essay in geopoetry." Presumably Hess was not comparing himself with that other "great poet," Alfred Wegener, for he neither cited nor mentioned Wegener. Hess wrote that "In order not to travel any further into the realm of fantasy than is absolutely necessary I shall hold as closely as possible to a uniformitarian approach. . . ." Further along in the paper, he noted that the work of another scientist suggests "that the water of the oceans may be very young, that oceans came into existence largely since the Paleozoic. This violates uniformitarianism, to which the writer is dedicated." (Uniformitarianism led Hess astray; the ocean basins are younger than the Paleozoic.) The implication of Hess's sidebar comments is that his musings need not threaten anyone, since they were only fanciful. Yet, should anyone care to take them seriously, that person could rest assured that uniformitarianism firmly underpinned the ideas.

Most of Hess's article is just what his title implied: the history of ocean basins. The diagrams that accompany the article depict seafloor spreading only secondarily. Their main purpose is to illustrate Hess's long-held views about the composition of the oceanic crust and upper mantle (which proved wrong). But his words did clearly convey the concept of seafloor spreading:

> The Mid-Atlantic Ridge is truly median because each side of the convecting cell is moving away from the crest at the same velocity, ca. 1 cm./yr. A continent will ride on convecting mantle until it reaches the downward-plunging limb of the cell. Because of its lower density it cannot be forced down, so that its leading edge is strongly deformed and thickened. . . .

Hess obliquely credited Holmes, writing: "Long ago Holmes suggested convection currents in the mantle to account for deformation of the earth's crust." There follow references to three specific authors "and many others," but not to Holmes, whose name does not appear among Hess's references. Hess did not cite Dietz, even though Dietz's seafloor spreading article appeared seventeen months earlier.

In the eyes of geologists, Hess would have had a distinguished career without plate tectonics. His "History of the Ocean Basins" is provocative and to his credit. But Menard may be the most objective witness to the conception of seafloor spreading. He had read both Hess's preprint and Dietz's manuscript and was in communication with both men. Menard believed that

> Only one person could have priority and that, by circulating his manuscript, Hess had it. We thought the similarities in the papers would prove embarrassing unless priority was openly ceded. These urgings [to Dietz] were despite the fact that I was certain, and said so, that Dietz had written his paper before he saw Hess's manuscript. I assumed that somehow Dietz had heard of Hess's ideas at some meeting or cocktail party and then forgotten about it. This is a common occurrence.

In other words, Menard and his colleagues took priority into their own hands, ceded it to Hess based on a preprint, and pushed Dietz into agreeing. Bowing to the urgings of his friends, Dietz added a "Note in Proof" to a 1962 paper, saying: "The writer's attention has been drawn to a preprint by H. H. Hess also suggesting a highly mobile sea floor. Full credit of priority is to be accorded him for any merit which this suggestion has." He went on to repeat his concession in three other papers. Dietz wound up with credit only for coining the phrase "seafloor spreading," but that turned out to mean little since "plate tectonics" shortly superseded it.

Hess may have discussed his ideas with Dietz, who may have forgotten them, as Menard suggests. If he had remembered, Dietz might have felt that the conversation was one of many and deserved no special recognition. In the late 1950s and early 1960s, wherever they gathered, geologists discussed many wild ideas about a contracting or an expanding earth, convection, the geophysics of the seafloor, wandering poles, and so forth. Hess later said the discussion took place; Dietz did not remember. Menard agreed with Dietz but pressured him into conceding anyway.

Those who wish to cede priority for seafloor spreading to Hess based on a preprint, to be consistent, must cede priority for the Vine-Matthews hypothesis to Morley. After all, before Vine and Matthews submitted their paper connecting seafloor spreading and magnetic reversals, Morley submitted the same idea as a letter to *Nature*. The editors rejected Morley's letter, but ought not have done so since a few months later they published the same idea by Vine and Matthews.

Scientific priority cannot go to the person who first has an idea because, as the Dietz-Hess controversy shows, that no one can divine. In this less than perfect world, priority must go to the first in print, the only fact in the matter that we can reliably establish. To base priority on casual conversation and preprints would allow anyone to have a claim and destroy the concept of scientific priority.

Thus, between the two of them, it is clear that, despite Dietz's several concessions, following the standard procedures and traditions of science it is he, and not Hess, who deserves priority for seafloor spreading. Settling this question was never really up to Dietz or Hess, whose self-interest is obvious, nor to Menard and the other scientists involved at the time, but to the larger scientific community and now, decades later, to historians of science.

But establishing Dietz's priority over Hess takes us only halfway. The remaining question is whether Dietz deserves credit over Holmes.

Naomi Oreskes, author of the most objective and comprehensive treatment of the reasons for the rejection of continental drift, writes: "It was Dietz, in 1961, who argued for the mechanism now accepted today—the mechanism that Holmes proposed in 1944—that oceanic crust is generated by suboceanic intrusion and submarine eruption of basaltic lava." Menard, at the center of the controversy, wrote, "All that is different between Holmes's hypothesis and sea-floor spreading is the median rift." (Which at the time Holmes wrote had not been discovered.) In his 1929 paper in the *Transactions of the Geological Society of Glasgow,* Holmes presented the classic figure shown on page 103, and wrote:

> The circulation due to unequal heating of the substratum would be a system of ascending currents somewhere within a continental region, spreading out at the top in all directions toward the cooler peripheral regions. The downward currents would be strongest beyond the continental edges. Where the ascending currents turn over, the opposing shears and the resulting flowage in the crust would produce a stretched region or a disruptive basin which would subside between the main blocks. If the latter could be carried apart on the backs of the currents, the intervening geosyncline would develop into a new

> oceanic region. . . . Where two currents meet and turn downward, the crust above the zone of contact will be thrown into powerful compression.

In the first version of his textbook, published in 1944, Holmes wrote:

> The currents have become sufficiently vigorous to drag the two halves of the original continent apart, with consequent mountain building in front where the currents are descending, and ocean floor development on the site of the gap, where the currents are ascending.
>
> Currents flowing horizontally beneath the crust would inevitably carry the continents along with them, provided that the enormous frontal resistance could be overcome.
>
> Most of the basaltic magma would naturally rise with the ascending currents of the main convectional systems until it reached the torn and outstretched crust of the disruptive basins left behind the advancing continents. . . . There it would escape through innumerable fissures, spreading out as sheet-like intrusions within the crust, and as submarine lava flows over its surface.

Both Dietz and Hess criticize Wegener for one alleged flaw: as Dietz put it: "A principal objection to Wegener's continental drift hypothesis was that it was physically impossible for a continent to 'sail like a ship' through the sima [basaltic crust]. Seafloor spreading obviates this difficulty: continents never move through the sima—they either move along with it or stand still while the sima shears beneath them." Hess made the same point: "A more acceptable mechanism is derived for continental drift whereby continents ride passively on convecting mantle instead of having to plow through oceanic crust." Neither appeared to know that Holmes had disposed of this red herring in 1944: "To sum up: during large-scale convective circulation the basaltic layer becomes a kind of endless traveling belt on the top of which a continent can be carried along, until it comes to rest when its advancing front reaches the place where the belt turns downwards and disappears into the earth."

In any event, there is no need to interpret Holmes's clear 1944 prose: upwelling convection cells in the mantle tear the crust apart and carry the rifted continents along to the margins of the ocean basins, where they become caught on down-welling cells and build mountains. Upwelling magma fills the crack and generates new ocean crust. If this is not what Holmes said, then words have no meaning.

But if the case is this clear, why did Holmes not receive the credit? The reasons may be several, but three stand out.

First, to give priority to Holmes would have done away with the argument that drift was rejected for lack of mechanism—by 1966, everyone had adopted Holmes's thirty-year-old mechanism: convection. Second, for Holmes to receive credit would have meant that the individual scientists who gave birth to seafloor spreading were doctrinaire and had ignored the voluminous geological evidence for continental drift. As late as the end of the 1950s—and even into the 1960s—most of them were on record as opposing drift. A new discovery—seafloor spreading—focused attention on the present rather than the distant past. Who could be faulted for having been less inventive than a new hero, Hess? Third, not only did Dietz cede priority, Hess demanded it, and no geologist of his era had more influence. In 1963, Hess wrote to a colleague: "In November 1960 I explained to Dietz in rather great detail the ideas in my paper. He rushed off and had these ideas published. There was no hint of any acknowledgement to me. Being as generous as I can I might suppose he forgot the source of 'his' ideas. But I believe this is a bit too generous."

In 1968, Arthur Meyerhoff, who opposed plate tectonics, published a paper in which he claimed priority for seafloor spreading for Arthur Holmes. Both Dietz and Hess responded briefly. Dietz once again ceded to Hess "full credit for the concept." Hess put his own priority on the record once and for all: "The cogent term 'sea-floor spreading,' which so nicely summed up my concept, was coined by Dietz after he and I had discussed the proposition at length in 1960."

In their responses to Meyerhoff, Dietz does not cite any writings by Holmes, and Hess cites only Holmes's 1929 paper. Both fault Holmes for only stretching and thinning the seafloor, not knowing or acknowledging that in his 1944 textbook, Holmes had ripped it asunder. From their papers, it is hard to escape the conclusion that neither Dietz nor Hess had read Holmes's textbook, and therefore both were unaware of the extent to which he had anticipated seafloor spreading. Dietz and Hess could hardly give credit to a book and author they had not read; though after Meyerhoff cited the book, reading it would have been the least they could have been expected to do.

Regardless of who should get credit for it, the test of a theory is its longevity and its ability to spawn new areas of research, to raise new questions and to suggest new ways of finding answers. Evolution is undoubtedly the greatest scientific theory extant, having single-handedly rationalized biology and having thwarted fanatical attempts to prove it false for most of 150 years. Paraphrasing what Theodosius Dobzhansky said about evolution, little in geology makes sense except in light of plate tectonics. It is the global theory that the greats of geology sought in vain.

PART III

CHANCE

12

Glimpses of the Moon

One of the great obstacles to progress is not ignorance, but the illusion of knowledge.

—DANIEL BOORSTIN

Most Fundamental Process

Surely the moon is the most scrutinized object in the heavens. While the sun is too bright to view directly, the irresistible moon—serene, lovely, and mysterious—demands our gaze and contemplation. Our heavenly companion has inspired poets and lovers, and puzzled scientists. Until the last two decades, the moon's origin proved so elusive that scientists joked that they had proved that she does not exist.

Galileo was the first to observe the moon through a telescope. He saw mountains and circular depressions and understood that the moon is not a perfect heavenly orb, as the Greeks had thought, but is a rocky world, with features and a history. Over the centuries since Galileo, many amateur astronomers, and a few professionals (who usually preferred to study stars and galaxies), turned their telescopes on the moon.

They reported astounding findings: coal dust, coral reefs, asphalt, insects, dust storms, snow, ice, glaciers, vegetation, people (Selenites), and atomic bomb craters. Many noted that while they were observing the moon, or in between their observations, something changed. They claimed to have seen entire craters appear and others disappear. Bright flashes and reddish patches materialized, as if from volcanic eruptions. Fog, mist, and haze (on the moon!) sometimes made the seeing difficult.

The year 1964, the four hundredth anniversary of Galileo's birth, was a time of ferment not only in geology but in lunar studies. Four different missions aimed to send, first, spacecraft and then men to the moon. The first, Ranger, would crash-land a camera-bearing spacecraft on the moon and send back close-up photographs. In anticipation of the information that was about to arrive, scientists put on two conferences. "The Dynamics of the Earth-Moon System," held at the Institute for Space Studies in New York City in January 1964, focused on its title subject but also included several talks on the origin of the moon. It apparently was the first conference to do so. The New York Academy of Sciences convened the second symposium, on "The Geological Problems in Lunar Research," which dealt principally with the moon's surface features. Not a single paper at the Earth-Moon System conference considered any of the topics taken up at the Geological Problems conference, and vice versa.

Minds at the two conferences must have been concentrated, as from the fear of imminent hanging, by the knowledge that previously untestable speculations would soon confront not only close-up photographs but, if all went well, actual samples. No longer would the moon be an inaccessible and unknowable object in space: a moon rock would soon sit on a laboratory bench. That some scientists, if not most or all, would turn out to be wrong was a certainty. In 1964, scientists disagreed not only over the three principal theories for the origin of the moon, but on much else: whether volcanism or meteorite impact created lunar craters; whether the surface of the moon is smooth or rough;

whether the moon is covered with a layer of dust so thick that anyone stepping on to its surface would sink like a stone; whether the moon formed hot or cold; whether it contains ice.

Over the decades before the Apollo missions, none of the three theories for the moon's origin had been able to win a majority of support. Indeed, each theory seemed to have at least one serious, if not fatal, flaw.

The "spouse" (capture) theory held that the moon formed in the outer part of the solar system, and as it lost energy, spiraled inward until the earth's gravity caught it. The "sibling" (double-planet or co-accretion) theory proposed that the moon and the earth formed at the same time and place, and by the same process. If so, the origin of the moon was bound to the origin of the earth: explain one and you have explained the other.

George Howard Darwin, son of Charles, proposed the third, the "daughter" (fission) theory and, as we saw in Part I, used it to calculate the age of the earth. He suggested that the pull of the sun's gravity on a rapidly spinning proto-earth raised higher and higher tidal bulges, until finally a piece broke off to become the moon. Darwin's theory followed from the nebular hypothesis for the origin of the solar system, in which a rapidly rotating sun flung off blobs that became the planets.

The three theories waxed and waned over the decades as new methods and data became available, but the flaws in each stubbornly refused to yield. The capture model demanded that the passing moon be in just the right orbit and be braked just enough so as neither to collide with the earth nor shoot past it, a vital requirement that confronts each returning spacecraft. And while astronomers recognized that Earth's moon is unusual in being so large compared to its planet, they preferred theories that might apply to all the satellites. Capture seemed ad hoc, which is always suspect.

The double-planet theory, unlike capture, had an advantage because it did not appeal to a *deus ex machina.* On the other hand, it failed to explain the large angular momentum of the Earth-Moon system. (Angular momentum is the measure of the motion of objects in curved paths, including both their rotation and their orbital motion. The angular momentum of the Earth–Moon system is the sum of the rotational spin of each body plus the orbital motion of the moon around the earth.) Calculations show that if both the earth and the moon evolved as co-planets from a primordial dust cloud, the combined spin of the pair would be much less than observed.

The double-planet theory also fails to explain the low density of the moon: 3.3 grams/cubic centimeter compared to 5.5 for the earth. Sibling planets should hardly show such a large difference in a fundamental property. On the other hand, 3.3 gm/cc is about the density of the earth's mantle, suggesting that if the moon came from the earth, as in the fission theory, it might have come from the mantle.

In spite of Sir Harold Jeffreys having pronounced fission impossible, the idea persisted. But angular momentum eventually did in fission as well. To fling off a piece of itself, the earth would have to have been spinning even more rapidly than today, with no obvious way of getting rid of the excess spin once its job had been done.

In contrast to the debate over the origin of the moon, until the 1960s, the origin of lunar craters had been about as uncontroversial as matters get in geology. For well over a century authorities had agreed that lunar craters, and indeed nearly every feature of the moon's surface, are volcanic. Presenters at the Geological Problems conference concluded that few if any craters form by the only rival method, meteorite impact. In an opening paper, one of the convenors, geologists Jack Green, a longtime proponent of lunar volcanism, criticized those who over the years had made overly dogmatic statements about the moon. But no sooner had he adopted this reasonable stance than he abandoned it, writing, "Meteoritic impact is considered to be a trivial process in affecting both the genesis and development of almost all major lunar surface structures."

In the years after the two conferences, the thousands of superb photographs of the moon and hundreds of precious samples returned raised more questions about the moon's origin than they answered. Instead of confirming one of the three theories, the new evidence weakened each. The moon appeared to be neither a simple spouse, sibling, nor child of the earth. But at the next conference on the moon's origin, held in 1984 at Kona, Hawaii, to the surprise of the convenors, paper after paper independently endorsed a new mechanism, one that combined the best features of each of the three classic theories. One after another, the speakers converged on the idea that a planet the size of Mars had struck the young earth a glancing but terrible blow, tilting our planet on its side and ejecting a mass of debris that settled into orbit and coalesced to become the moon.

Nearly two decades after the 1984 conference, the giant impact theory remains robust. Impact, and not volcanism, created almost all the moon's craters and surface features. To be sure, basalt fills the lunar basins and produces dark spots here and there, but impact produced the basins themselves.

Impact has caused or affected everything from the origin of the planets and moons to, at least once, the course of evolution. Indeed, the increased recognition of the role of meteorite impact in the birth and life of the solar system constitutes a scientific revolution as profound as any in the earth sciences. But the impact revolution, like those led by Copernicus, Hutton, and Darwin, bears a cost to our hubris. Impact is random. Chance alone dictates when, where, and with what energy meteorites strike. In this view of the history of the solar system, almost everything important, including our very existence, is due not to destiny but to chance.

Selenographers

Besides the dark and light areas of the moon's face, which Johannes Kepler called the maria and the terrae respectively, the moon's circular craters are its most obvious features. As soon as early telescopes rendered the craters visible, scholars began to speculate about their origin. The brilliant seventeenth-century polymath, Robert Hooke (1635–1703) Secretary of the Royal Society, made experimental craters by dropping musket balls into wet clay. He produced pits that resembled lunar craters, but since he and his contemporaries believed space to be empty, he could not imagine from where the cosmic analogs of the musket balls might have come. In his *Micrographia* (1665), Hooke concluded that the craters are of internal origin, like volcanoes on earth. Hooke was the first of a long line of observers to get the moon and its craters wrong. To get them right would take three hundred years and a number of manned voyages.

Astronomers, geologists, and "selenographers," as the early students of the moon called themselves, with only a handful of exceptions agreed with Hooke: the moon's surface is volcanic. The accepted author of the contrary theory—that meteorite impact created the craters—was a German selenographer named Franz Gruithuisen (1774–1852). Unfortunately, confidence in Gruithuisen's science was undercut by his claim that he had observed people, animals, and cultivated fields on the moon—even a fortress and a temple at which the "'lunites' worshipped the stars." Even the great British astronomer Sir William Herschel (1738–1822) averred in 1780 that it was a near certainty that people inhabited the moon.

Showing that it is possible to improve almost anything, in August 1835, the *New York Sun* began a series of articles describing the observations of the moon allegedly made by Sir John Herschel (1792–1871), Sir William's equally illustrious son. Sir John's assistant signed the stories, the astronomer himself being incommunicado in remote South Africa. The articles described how Sir John had seen "sheep, pygmy zebra, unicorn roaming lunar grasslands, as well as bipedal winged creatures called manbats." It turned out that an inventive reporter for the *Sun,* desirous of selling more papers, wrote these exciting reports.

In 1849, the real Sir John concluded that lunar craters are close analogs of the volcanoes of France and Italy. He maintained this position for the next fifty years in the various editions of his influential textbook; most other authors followed his lead. Perhaps the most authoritative voice was James Dwight Dana of Yale, who had also declared himself on continental permanence and geosynclinal mountains. In his famous *Manual of Geology,* which went through four editions, Dana wrote that the moon "presents to the telescope a surface covered with the craters of volcanoes. . . . The principles exemplified on the earth are but repeated in her satellite."

One reason that the earliest students of the moon did not conceive of impact as a realistic hypothesis is that, like Hooke, they could not imagine what might have fallen to create the craters. But after the descent of thousands of meteorites on the French village of L'Aigle in 1803, witnessed by hundreds, and after the discovery of asteroids (named by Sir William), impact became at least a theoretical possibility.

Closest to a Saint

By far the most distinguished scientist to credit impact as the cause of lunar craters was Grove Karl Gilbert (1843–1918), chief geologist of the U.S. Geological Survey. Gilbert has been described as "surely one of the greatest geologists who ever lived" and the "closest thing to a saint that American science has yet produced." He was a geomorphologist: a student of the origin and development of land forms. In 1891, Gilbert attended a meeting of the American Association for the Advancement of Science in Washington, D.C., where he heard mineralogist Arthur E. Foote describe a "circular elevation . . . occupied by a cavity" close to the Canyon Diablo trading post in northern Arizona. Nearby were scattered a number of iron meteorites, one of which contained diamonds. Whereas most in the audience may have been interested in the diamonds, the "so-called crater . . . [which] was like the depressions on the surface of the moon produced by impact of an enormous meteoritic mass" intrigued the scholarly Gilbert.

Gilbert had already been thinking about the connection between craters and impact, even wondering whether accretion of meteorites might not have constructed the planets. He decided to visit the Arizona cavity and see for himself. Gilbert was an exponent of the proper methodology of science and of the principle of multiple working hypotheses. He wrote: "The man who can produce but one [hypothesis] cherishes and champions that one as his own, and is blind to its faults. With such men, the testing of alternative hypotheses is accomplished only through controversy. Crucial observations are warped by prejudice, and the triumph of truth is delayed." He noted that no conclusion "is so sure that it can not be called into question by a newly discovered fact." Gilbert's words would ring ironically prophetic as his failure to live up to them delayed the triumph of truth by more than half a century.

In contrast to Gilbert, Foote's interest in the so-called Arizona crater, like others who would become obsessed by it, was not scientific but pecuniary. In a hole in the ground in Arizona, some thought they saw the end of the rainbow and a pot of gold.

Gilbert's own thinking about lunar craters, together with the report of a colleague whom he sent to Arizona to study Coon Mountain (as the crater was at first misleadingly named), led him to conclude that it could have formed in one of two ways: steam might have erupted from underground, producing a cavity but leaving no telltale igneous rock behind; or a meteorite might have fallen. In his logical way, Gilbert devised tests that would discriminate between the two possibilities: First, if an underground explosion had blasted out the material that once resided in the crater, that same material should now be found lying on the surface around the rim of the crater and it should have the same volume as the crater it had once filled. On the other hand, if a "star entered the hole the hole was partly filled thereby and the remaining hollow must be less in volume than the rim." Second, if an iron meteorite created the crater, the floor ought to conceal the "buried star," whose magnetism would betray its presence.

In 1891, Gilbert and his party set out for Flagstaff, the town nearest to the crater and the future home of the Astrogeology Branch of the U.S. Geological Survey. Gilbert thought he already knew what his tests would show: "the presumption was in favor of the theory ascribing the crater to a falling star, because the theory explained, as its rival did not, the close association of the crater with the shower of celestial iron." Once on site, Gilbert and his party conducted careful geological, topographic, and magnetic surveys. Contrary to his presumption, the volume of material apparently ejected from the crater turned out to be the same as the volume of the crater itself, satisfying his first test and suggesting an internal origin. We now know, as Gilbert could not, that the identity of volumes is a coincidence. He and his party found neither igneous rocks nor anomalous magnetism, satisfying the second test.

Having erected a petard of scientific logic and observation, Gilbert now could not escape hoisting himself atop it. In 1895, he presented his conclusions in a paper entitled "The Origin of Hypotheses, Illustrated by the Discussion of a Topographic Problem." The problem was Coon Mountain. Gilbert delivered the paper from his bully pulpit as president of the Geological Society of Washington, D.C. Although he had journeyed to Arizona expecting to find evidence of impact, and although he had maintained the two working hypotheses, given the way he had framed the question, Gilbert could come to only one conclusion: steam volcanism, and not impact, had created the crater.

His inexorable logic caused Gilbert to have to attribute the scores of meteorites scattered around the crater to coincidence. Such an illogical result ought to have led Gilbert to question his assumptions, and near the end of his talk, he appeared ready to do so. He reported that a colleague had found that each of the Canyon Diablo iron meteorites was slightly different, as though each had come from a different parent. Thus, the meteorites might have been included in some larger, non-magnetic object, "like plums in an astral pudding," leaving the hypothetical buried star with much lower magnetism than he had originally conjectured. In that case, the absence of magnetism might not disprove the impact theory after all. Not only that, but perhaps some of the rocks had been "condensed by shock so as to occupy less space," which would mean that the test based on the volume of ejecta could also mislead. Gilbert noted that these ideas were "eminently pertinent to the study of the crater and will find appropriate place in any comprehensive discussion of its origin." Yet for the remaining twenty-three years of his life, Gilbert refused to find any place for such ideas and failed even to acknowledge new findings that went against his conclusions.

Prompted by his study of Coon Mountain, for eighteen nights in the late summer and fall of 1892, Gilbert observed the moon with the telescope of the Naval Observatory in Washington. Anticipating the latter-day "Golden Fleece" awards of Senator William Proxmire, given to alleged government-supported research boondoggles, one congressman in Gilbert's day claimed that "So useless has the Survey become that one of its most distinguished members has no better way to employ his time than to sit up all night gaping at the Moon." But never before had a scientist so skilled at observing landforms studied the moon.

Gilbert observed that the floors of lunar craters, unlike those of terrestrial volcanoes, lie below the level of the surrounding plain. Even the summits of the central peaks found in the larger lunar craters are often below the level of the plain. He saw that some lunar craters are many times the size of the largest terrestrial volcanoes. The lighter-colored "rays" that radiate outward from some craters for scores of kilometers, Gilbert interpreted as splashes from impacts. He concluded: "The volcanic theory as a whole, is therefore rejected."

Not every observation was consistent with meteorite impact,

Gilbert recognized. The circularity of lunar craters puzzled him most. Since objects flying in from space would strike the surface of the moon at random angles, some of them barely glancing, at least some lunar craters should be oval-shaped. Gilbert proposed that meteorites had not arrived at oblique angles, but had dropped vertically from a ring of debris. He postulated that, early in the history of the solar system, the earth was surrounded by a cloud of small, solid objects which "gradually coalesced" to form the moon. "The lunar craters are the scars produced by the collision of those . . . moonlets which at last surrendered their individuality." The last few objects—his moonlets—had descended vertically, not at random angles, and therefore had created circular craters. While this idea was wrong in detail, it did presage the currently accepted theory for the origin of the moon itself.

Gilbert presented his work in a paper called "The Moon's Face." Curiously, not only does the paper not mention Coon Mountain, his "Origin of Hypotheses" paper does not mention "The Moon's Face." It is as though in his mind Gilbert completely separated the lunar craters, which he concluded were due to impact, from the earthly ones, which he concluded were not. He got the origin of Coon Mountain wrong and almost everyone agreed with him. Using the same logical approach, he got the origin of lunar craters right, but almost no one agreed with him.

In the two decades following the "Origin" paper, Meteor Crater, as people came to call Coon Mountain, gave rise to much new and significant evidence. Yet, betraying his professed philosophy, for the rest of his life Gilbert never wrote or said a single word publicly about Meteor Crater. His reaction to repeated entreaties to do so was simply to ignore them.

The effect on the U.S. Geological Survey was to put research on meteorite impact off limits. The long shadow that Gilbert cast over impact studies stretched to 1959, sixty-four years after his Washington address on the origin of hypotheses, when another Survey geologist falsified Gilbert's conclusion about Meteor Crater. In part for his work in showing that impact created Meteor Crater, and with no small irony, in 1983 Eugene Shoemaker won the G. K. Gilbert Award of the Geological Society of America.

Obsession

On a warm Tucson night in 1902, a pair of men smoked and swapped stories. Samuel Holsinger had traveled widely in Arizona Territory and had heard the tale of Coon Mountain. He passed the story on to his evening companion, mining engineer Daniel Moreau Barringer (1860–1929). On his return home, Barringer told his wife about the crater and later said that "ridiculous as it might seem, I had made up my mind to inquire more fully into the matter." In spite of Gilbert's conclusion, of which he was well aware, Barringer formed a lifelong conviction that the meteorite was still present under the crater and that he could find and mine it. Before he could pursue his dream, Barringer had to acquire rights to the crater, which he did by stealthily staking a claim on the government-owned land. Barringer was to spend a fortune, and nearly all of his remaining years, energy, and influence, in a vain quest for the elusive meteorite.

Barringer devoted full time to promoting the incipient mine, writing hundreds of letters to scientists, engineers, politicians, and government employees. In March 1908, he wrote to the director of the U.S. Geological Survey, Charles Doolittle Walcott, to bring him up to date on the new research that Barringer had sponsored at the crater, and which he at least believed greatly strengthened the case for impact. He tried to be at his most diplomatic, saying that he appreciated that the Survey "did not have the advantage of the information which has been furnished by our extensive development work. . . . " But for Barringer to be right, Gilbert would have to be wrong. Gilbert had never allowed that possibility even to be considered. Walcott never replied.

Twenty years later, Barringer tried again in a long letter to the new director of the Survey, George Otis Smith. By now, Barringer believed that new findings had proved that impact had created the crater, as did almost everyone who had reviewed the evidence. Gilbert was long in his grave. Barringer beseeched Smith to "state in a short article" that Gilbert and the Survey did not deserve credit, "even by inference," for the impact theory. In reply, Smith cited Gilbert's paper of 1896, in which "Dr. Gilbert specifically assumes responsibility for the meteor impact theory of the origin of the features. . . ." Gilbert had not had the advantage of the new evidence "supporting one or another of the theories which he discussed in so searching and fruitful a way," wrote Smith. Barringer fumed, "Surely it is a new thought that the upholder of an incorrect theory is entitled to credit for the correct one, simply because he attempted to refute it."

Unbeknownst to Barringer, and indeed to most of the scientific world, a young Estonian astronomer named Ernst Jules Opik (1893–1985), in an obscure scientific paper written in 1916, had exposed the futility of Barringer's obsession. Opik would become famous for his studies of asteroids and comets, as well as for his prediction of craters on Mars and the existence of the giant Oort cloud of comets. He deduced why impact craters are circular, the feature that had puzzled Gilbert and other observers: The energy of a moving object is one half its mass times the square of its velocity. Even a small body, traveling at high speed, contains a great deal of energy. A meteorite colliding with another body at cosmic velocities, in the range of 20–70 km/sec (50,000–150,000 mph), contains such enormous energy that it explodes. Opik and later others showed that impact produces a crater much larger than the meteorite itself; that the resulting explosion causes craters to be circular regardless of the angle of impact; and that most of the impacting meteorite is vaporized, leaving at most a few meteorite fragments, like those found around Meteor Crater.

Opik's conclusion came to be generally accepted, though not by Barringer, as a result of the work of the astronomer Forest Ray

Moulton, Thomas Chamberlin's colleague. Moulton wrote to Barringer that he had discovered during World War I that "the impacts made by oblique projectiles are essentially circular." Moulton issued reports that demonstrated conclusively that impact created Meteor Crater and that the resulting explosion blasted the meteorite to pieces. In one report, he wagered that "if more than 500,000 tons of meteorite are found, take a long hearty laugh at my expense and call on me to put up a dinner at the crater. . . ."

Barringer died in 1929, and no one ever claimed Moulton's bet. Over the next few years, the investors who had joined Barringer accepted the inevitable: impact created Meteor Crater; to search for the almost completely vaporized meteorite would be futile. The Meteor Crater impactor—Patterson's Canyon Diablo—weighed about 300,000 tons and disappeared in a 15-megaton event, about the magnitude of the Hiroshima atomic bomb explosion. Moulton's bet would forever be secure.

Gilbert concluded that impact produced the moon's craters but that volcanism, in the form of a steam explosion, produced Meteor Crater. Most geologists accepted a volcanic origin for both. Only a handful believed that impact created not only lunar craters, not only Meteor Crater, but many other terrestrial craters as well.

Drifters

In reading the literature on lunar craters, one is at first surprised, and on further thought not so surprised, to find two names familiar from another controversial theory, continental drift. The initial surprise comes because we expect scientists, as they do today, to have worked in narrow disciplinary niches. We hardly expect one who writes on continental drift to comment on something as far afield as meteorite impact on the moon. But always a few scientists cannot be categorized. These risk takers are often wrong, but now and again they see farther.

During 1918 and 1919, while Alfred Wegener was a dozent at the University of Marburg, perhaps in between editing his book on drift, he turned his attention to the moon. Like Hooke, Gilbert, and probably many others, Wegener cast solid projectiles into a mud bath to simulate lunar craters. His experimental pits closely resembled those on the moon, convincing him that the impact hypothesis, "which is most rarely represented by the experts, on close examination proves to be the most promising." He simulated beautiful lunar rays, like those we see emerging from the crater Copernicus, as well as the central peaks that mark the large lunar craters.

But if impact created the moon's craters, why are there so few craters on Earth? Wegener concluded that the terrestrial impactors had fallen into a molten earth and disappeared, whereas those that struck the moon did so after it had frozen, allowing it to retain the scars of impact. He compared terrestrial and lunar craters and noted that "The forms are fundamentally different; therefore, the origins also should be different." "The contradiction is so flagrant," Wegener remarked, "that the next generation will only laugh at our desperate experiments with which we try to establish an equality between the Moon and Earth."

Since boyhood, the moon had fascinated Robert Dietz. He proposed writing his Ph.D. thesis on the moon's surface features, but his professors turned him down, joking that no one would be able to check his fieldwork. But Dietz continued to wonder about craters. His interest was caught by some unusual rock structures exposed in a quarry etched into the flat cornfields of northwest Indiana near the small town of Kentland, not too far from Urbana, Illinois, where his university was located. One day in 1943, it occurred to Dietz that "the disrupted nest of lower Paleozoic strata in the Kentland quarry might be . . . an asteroidal impact scar." With the typical "can-do" attitude that characterized young Americans during the war, while engaged in flight training one day, Dietz decided to do some fieldwork after all. He landed his plane at the airfield nearest to Kentland and hitchhiked to the quarry. There he found structures known as shatter cones, in which grooves in a rock radiate outward like the feathers of a badminton shuttlecock or a horse's tail.

Shatter cones subsequently turned up at other puzzling geologic structures, where rocks have been broken into a series of concentric rings, like pond ripples turned to stone. The leading expert on these cryptovolcanic structures was the anti-drifter Walter Bucher. He and others were convinced that, as the name suggested, a volcanic force directed from below had produced the concentric fractures. Mysteriously, this force left no evidence other than the faulted rocks—having committed the perfect geologic crime, this silent killer vanished without a trace. While this interpretation may not have been strictly uniformitarian, at least it appealed to a known geologic process, volcanism, rather than to the then unknown one of rocks falling from the sky. But Dietz was never one to accept the doctrinaire answer. The archetypal outsider, in more than forty years of doing science he never received a research grant.

After he had mentally rotated the broken and folded rocks at Kentland back to their original horizontal positions, Dietz found that the apices of the shatter cones pointed skyward. He concluded that the forces that created the Kentland structure had been "cosmic rather than volcanic." Dietz and others later found shatter cones at many other cryptovolcanic structures, including the two largest and oldest, the Vredefort Ring in South Africa and the Sudbury structure in Ontario, both now accepted as due to impact.

Immediately after the war, while supervising the oceanographic research on Admiral Richard E. Byrd's last Antarctic expedition, Dietz's fertile mind wandered again to the moon's surface features. He wrote up his thoughts and submitted them in a paper to the *Journal of Geology,* whose editor was none other than Rollin T. Chamberlin (son of the great Thomas), who had himself just written a paper "on the tectonic [i.e., non-impact] origin of the lunar geomorphology." The

Journal published Dietz's paper in 1946; it was the first, he said, to suggest an impact origin for lunar craters since Gilbert's paper in 1895. Dietz methodically confirmed the distinctive properties of lunar craters that Gilbert had noted, and added others.

The case Dietz made for impact as the cause of lunar craters was overwhelming, yet no one paid any attention. To move geologists to change their minds about the origin of the moon's craters, something more was required than observation and qualitative conclusions.

13

Moonlighting

In research self-satisfaction is death! Doubt motivates progress, but it is painful to endure.

—Jacques Monod

A Field to Himself

In early 1941, astrophysicist Ralph B. Baldwin (b. 1912) was an instructor at Northwestern University and about to become a father. In order to support his soon-to-be-increased family, he began to moonlight (never was the euphemism more appropriate) by giving lectures at Chicago's Adler Planetarium at $4 a ticket. When he arrived early, Baldwin wandered the halls, examining the Adler's exhibits and superb photographs of the moon. One in particular caught his eye: it showed a set of "long valleys with raised rims" that he had never seen mentioned in the literature. He noticed that these valleys "pointed backward toward Mare Imbrium."

His curiosity aroused, Baldwin searched for references to the strange valleys. Finally, he found a paper that attributed them to "the nearly tangential impact of a swarm of huge meteorites," a suggestion that at first did not strike him as reasonable. Further study of lunar photographs soon caused Baldwin to change his mind and to conclude that "the circular maria had been produced by gigantic explosions—only the impact of giant meteorites could supply the requisite energy." But this idea, he soon learned, was anathema; only a handful of professional astronomers had been willing to entertain meteorite impact as the cause of lunar craters. A few weeks later, Baldwin attended a colloquium with the most distinguished astronomers of his day. He soon discovered that none of them knew or cared much about the moon—his few weeks of study had left him "knowing far more about the moon than any of them." Baldwin realized that "nobody else was actively mining this lode" and that he had "a significant field almost to himself." He wrote up his ideas on meteorite impact on the moon and submitted them to leading journals of astronomy—all of which rejected his paper. Finally, *Popular Astronomy* published it. Annoyed by the dismissal of his ideas, Baldwin set out to prove his critics wrong. He would show that the moon was a worthy subject for study and that his theories were sound.

After spending the wartime years helping to develop the radio proximity fuze (an explosive ignition device used in bombs and artillery shells), Baldwin made his first major contribution to lunar geology in a book entitled *The Face of the Moon* (1949). In it he reported his measurements of the dimensions of 329 lunar craters. He found that on a logarithmic graph of depth versus diameter, the craters traced out a smooth curve—the wider the deeper—over a range of diameter of 150 times. What should have been the clincher was that explosion pits created by wartime mortar and bomb shells fell right on the lunar curve. Baldwin concluded that "The only reasonable interpretation of this curve is that the craters of the moon, vast and small, form a continuous sequence of explosion pits, each having been dug by a single blast.

No available source of energy is known other than that carried by meteorites."

He summed up: "To claim that the moon's craters are volcanic is tantamount to postulating an entirely new, entirely hypothetical mode of origin and to fly in the face of the fact that a known process is completely able to explain the vast majority of observed lunar features." In a sense, Baldwin attempted to extend uniformitarianism to the moon and to broaden the concept to include processes that no one has actually seen but which we can infer. But this was too much for most people to accept. In any event, in 1949 only a handful of scientists had any interest in the moon and its craters. To those who did, the matter appeared settled: lunar craters are volcanic.

Urey and Shoemaker

One who read *The Face of the Moon* was the American Nobel laureate and chemist Harold Urey. The book so inspired Urey that he made the origin of the moon and planets the focus of the remainder of his life's work. Another was a precocious graduate of the California Institute of Technology, Eugene Merle Shoemaker. When Baldwin's book appeared, Urey was at the University of Chicago and was already famous; Shoemaker was only twenty-one. Urey was often grandly wrong; Shoemaker was usually right. Both made major contributions to geology. To his credit, when Urey was wrong, he promptly admitted it and then modified his ideas to incorporate the new evidence. If anything, he may have been too ready to change his mind. But how much better that than a refusal to debate or to admit new evidence.

Urey read *The Face of the Moon* on a train trip to Canada, lined his University of Chicago office with lunar photographs, and became consumed with the moon. His interest was not so much in the origin of lunar craters—he accepted Baldwin's logical conclusion that they were due to impact—as in the origin of the moon itself. Urey knew that the moon bulges slightly toward the earth, perhaps due to centrifugal force and the pull of the earth's gravity. He speculated that the bulge had formed early in the moon's history and had been locked in. Since the bulge has not receded, the moon must not have isostasy; its interior must be rigid. A cold, rigid moon was likely to be primordial and thus to be a fossil from the earliest history of the solar system. By studying it, we could learn about the starting conditions of all the planets and moons.

Urey published his ideas in a book called *The Planets, Their Origin and Development* (1952), which was widely read and so persuasive that it caused many physical scientists to change the direction of their research. Baldwin begat Urey; Urey begat a host of superb scientists. Fortunately, parents vest not only their sins in their children, but their good deeds as well.

Shoemaker's remarkable career and tragic death have been well recorded elsewhere. If Arthur Holmes ranks as the greatest geologist of the first half of the twentieth century, Shoemaker is the obvious candidate for the title in the second. Both were consistently ahead of their time. To retain the respect of most of one's colleagues while being far out in front of them took not only great intellect but also unusual character and resolve.

At the age of only twenty, already with a master's degree from the California Institute of Technology, Shoemaker was at work in the remote Colorado Plateau. In the Cal Tech campus paper, forwarded to him in the field, and one of his few contacts with the outside world, he read that the Jet Propulsion Laboratory at Pasadena was conducting experiments with captured German V-2 rockets. Shoemaker made the kind of mental leap that would characterize much of his work: "Why, we're going to explore space, and I want to be part of it! The moon is made of rock, so geologists are the logical ones to go there—me, for example." But Addison's disease kept Shoemaker's feet, if not his mind, on the ground.

After his epiphany while reading the campus newspaper, perhaps the first time that genre has inspired such ambition, Shoemaker read Baldwin's *The Face of the Moon.* Though Urey, given his age, stature, and university tenure, was free to study whatever he wished, Shoemaker worked for the USGS. For half a century, the Survey had not allowed its employees to study terrestrial craters, much less lunar ones. As a way to pursue his interest in the moon while acceding to the boundaries established by the Survey, Shoemaker hit on the idea of studying the nearby Hopi Buttes, which he thought resembled lunar volcanic necks and chains of craters on the moon. Shoemaker justified this work by convincing his superiors that, as the volcanic material that was to become the buttes rose to the surface, it might have gathered up uranium deposits, a major focus of the Survey at the time. From there he moved on to study actual craters, produced not by meteorites or volcanoes, but by underground nuclear explosions at the Nevada Test Site.

Shoemaker soon became aware of the nearby Meteor Crater. Having as yet no contrary evidence, he initially followed Gilbert in believing it to be the product of a steam explosion. The descendants of Daniel Barringer still owned the crater and did not welcome a visitor from the hated Survey, especially if he were not ready to endorse a meteoritic origin for their crater. Shoemaker won them over by having a friend of the family vouch for him to the son of Daniel Barringer. Thus began Shoemaker's classic study of the former Coon Mountain.

In 1959, Shoemaker put to rest the notion that Meteor Crater is of terrestrial origin. He showed that impact is unlike a chemical explosion, in which a reaction causes material to vaporize rapidly. Rather, impact generates two shock waves that interact in a complex fashion to excavate the cavity and destroy the incoming meteorite. When the meteorite first makes contact with the surface, one shock wave moves back into the meteorite to engulf and blast it to smithereens, in the process destroying the dreams of meteorite miners. Ahead of the incoming meteorite, another shock wave races down into the incipient crater, compressing the target rocks and then ejecting them at great speed. Fragments of target rock thrown out on low trajectories land in the reverse order in which they departed, stacking upside down relative to their original layering—older above younger. Much of the nearby ejecta slumps back into the crater. Shoemaker showed that the rocks around Meteor Crater perfectly manifest this theoretical process. The *coup de grâce* to Gilbert's notion of a steam explosion was the discovery at Meteor Crater of two varieties of quartz that require pressures so high that they had previously been seen only in the laboratory.

Anti-drifter and anti-impactor Walter Bucher continued to hold out for a terrestrial origin of Meteor Crater and all cryptovolcanic structures. His last paper, delivered at the 1965 conference on "The Geological Problems in Lunar Research," was entitled "The Largest So-Called Meteorite Scars in Three Continents as Demonstrably Tied to Major Terrestrial Structures." Bucher claimed that the largest crypto-volcanic structures are associated with even larger geologic features. This association, if true, would mean that cryptovolcanic structures are not located randomly, which in turn would mean that they are not cosmic but terrestrial. But here Bucher proved to be wrong. According to scientist-author Ursula Marvin, just before he died, Bucher accompanied Shoemaker on a field trip to Meteor Crater and became persuaded of its impact origin, but did not have time say so publicly.

Rosetta Stone?

As a few scientists began to entertain the possibility that impact might have created lunar craters, attention turned to possible terrestrial analogs. In his 1949 manifesto, Baldwin cited ten authenticated terrestrial meteoritic craters. But, he wondered, "How many other gigantic meteorites have disappeared into the watery wastes? What myriads of meteoritic craters lie unseen on the surface or for ages hidden in stony crypts?" In answer to this poetic question, by 1960 geologists had identified 32 terrestrial impact craters; today the number stands at 160, with more discovered each year.

Thus, by the time of the 1964 conference on Geological Problems, just prior to the first missions to the moon, the argument that the earth contained no impact craters and therefore the moon, our close companion, could contain none either, had been falsified. Yet the conference speakers continued to attribute the moon's surface features to volcanism. This long-held position was about to collide with the hard evidence of close-up photographs and rocks from the moon.

Ardent volcanist Jack Green, a former student of Bucher, organized and chaired the conference. In his opening paper, Green praised the work of a mining geologist, Josiah Spurr, whose biographical memoir Green would write a few years later. In four volumes published privately during the 1940s, entitled *Geology Applied to Selenology*, Spurr spurned impact and saw instead in the photographs of the moon a "lunar grid": a set of linear features that ran N-S and E-W, like the lines of latitude and longitude on the earth. Since impact is random, if such linear features did exist on the moon, they would have been internal. Of course, the human mind creates lines of terrestrial latitude and longitude. Spurr's lunar grid proved equally imaginary.

At the moment when the first close-up photographs from Ranger began to emerge, a few scientists, like Green, held that volcanism had created nearly everything on the surface of the moon, including the large, fresh crater Copernicus. (As late as 1970, Green published a paper entitled "Copernicus as a Lunar Caldera.") Even the most zealous impactors, like Urey, Baldwin, and Shoemaker, accepted that many, if not most, of the smaller lunar features are volcanic. Most experts thought that tektites—the glassy, streamlined objects found scattered about the earth—had come from the moon. In that case, thousands of lunar specimens are already available in museum collections. Thomas Gold of Cornell announced that hundreds, perhaps thousands, of meters of fine dust, into which an ambling astronaut would sink like a stone, filled the maria: the large lunar basins. He thought the dust had settled from space and that electrostatic forces moved it around on the moon's surface. Like Spurr's grid, Gold's dust would "cost considerable time and money before it was finally disposed of."

After hundreds of years of viewing the moon from a distance that made any theory safe from a reality check, by the mid-1960s, the effective distance was about to close to zero. Surely, scientists would now be able to translate the hieroglyphics of lunar geology. Surely, the Rosetta stone of the solar system would now reveal its long-held secrets.

14

Voyages

I can state flatly that heavier than air flying machines are impossible.

—LORD KELVIN

First in Space

On October 4, 1957, the world awoke to the beep-beep of the 84-kilogram Russian satellite Sputnik sailing far overhead. *Homo sapiens* had left its home planet and taken its first step toward the stars. When Shoemaker heard the news at his field station in the Hopi Buttes, he responded, "But I'm not ready yet!" Neither was anyone else in America. The United States had no space program and, as would shortly become all too apparent, no reliable rockets with which to launch anything into space. As to the most obvious target of a space program—the moon—no one knew much about it. What they thought they knew would prove wrong.

A month after Sputnik, the U.S. sense of post–World War II complacency and dominance was further shattered when the Russians successfully launched another satellite, this one carrying the ill-fated dog Laika. Before America had barely awoken, the Russians appeared poised to send cosmonauts, if not a hydrogen bomb, over its head.

America responded by accelerating the existing work on rocketry that had begun at the Jet Propulsion Laboratory and elsewhere, and by creating the National Aeronautics and Space Administration (NASA). The U.S. Geological Survey, at Shoemaker's urging, established an Astrogeology Branch and installed him as its head. The new branch began by tackling three problems. The first, the origin of tektites, was to prove a red herring. At the end of the 1950s, most scientists thought that impact had splashed the glassy tektites off the moon with enough velocity to send them to Earth. As they entered the earth's atmosphere, the tektites melted and took on their streamlined shapes. Thus, pieces of the moon might already be present on the earth. If so, scientists needed to study them before spacecraft and astronauts went to the moon to collect specimens.

The second effort of the new Astrogeology Branch was to use the techniques of stratigraphy—the study of the sequence of layered rocks—to create a geologic map of the moon. Every feature shown on a lunar photograph that cuts, or lies atop, another feature reveals which of the two is older. For example, a crater etched into a mare basin must be younger than the material that fills the basin. When one crater cuts the rim of another, the one doing the cutting must be younger. A recognizable lunar surface unit, or formation, that overlies another unit must be younger, and so on. These common-sense concepts, which go back to Steno's law of superposition, allowed a detailed lunar stratigraphy to be developed. One of its most important findings was that the maria basins were created long before they were filled, like a bowl that exists before the soup that later is poured into it.

The third project of the fledgling branch was to simulate craters by shooting hypervelocity projectiles into rock targets. In 1960, Don Gault fired aluminum spheres into the Kaibab Limestone from Meteor Crater and produced perfect shatter cones.

On May 25, 1961, came the riveting speech in which John Fitzgerald Kennedy challenged his nation to send men to the moon and return them safely before the decade was out. The moon was not only the most obvious target, it was the only sufficiently dramatic goal that NASA could achieve in a reasonable time and at an affordable cost. Mars was too far away: a journey there would require at least six months in space, with who knows what toll on the astronauts. To find the nation's sea legs on JFK's new ocean of space, it made sense to take a short voyage first.

Almost two years before JFK's speech, the Russians had sent Luna 2 crashing into the rim of the crater Autolycus. Surely they intended more missions to the moon, some of which would likely land cosmonauts on its surface. But by 1961, they had not yet done so, leaving the door open for an American to be the first to walk on the moon.

But simply beating the Russians to the moon was not a sufficient reason, or at least not one that politicians could defend, to justify the expenditure of billions of U.S. dollars. A scientific reason was also needed, and scientists, led by Harold Urey, were glad to provide it. Although most astronomers had disdained study of the moon—Carl Sagan said it was boring—Urey had persuaded many that the moon is a deeply important object.

By 1965, though Kennedy had been slain by the great coward, Lee Harvey Oswald, the stage was set to fulfill JFK's challenge. The United States had broken out of a dismal pattern of launch failures and had assembled and trained a team of photogenic, crew-cut astronauts, each exemplifying the "right stuff." Shoemaker's Astrogeology Branch had mapped the near side of the moon in detail from photographs and identified possible landing sites. The United States was ready.

"It's Basalt, Isn't It?"

Before sending men to the moon, NASA had to learn whether the surface would bear the weight of a landing vehicle. The early Ranger missions, designed to return close-up photographs of likely landing spots just as the craft crashed into the moon, had begun in 1959, but had usually failed. Finally, in late July 1964, Ranger 7 struck only a few kilometers from its intended landing site on a ray of ejectra from Copernicus. The 4,300 photos telemetered back to Earth showed a monotonous surface covered with craters, but no lunites, manbats, or obvious volcanic features. The photos did show grid marks—earthbound technicians had superimposed them to provide reference points. Rocks rested in small indentations, suggesting that if Gold's dust were present, it formed only a thin layer.

The volcanists, or hot-mooners, typified by Spurr and Green, thought the moon had been and still was geologically active, just as is the earth. The impactors, or cold-mooners, typified by Urey, thought the moon had accumulated from cold particles elsewhere in the solar system and had wandered near the earth and been captured by it. Since then, only impact has affected the moon. The debate was important, because the mission planners thought that the ash and lava of a hot moon would likely bear the weight of a spacecraft and astronauts, whereas a dangerous layer of impact-generated debris might coat a cold moon.

Before samples of the moon were available, the only way to measure the age of its surface was to assume that impact created lunar craters and to estimate crater density. If one knew the number of craters on a mare surface and could estimate the rate of crater production, one could divide one by the other to obtain the age of the surface. William Hartmann, scientist, artist, science fiction author, and member of the Ranger Interpretation Team, took a crack at estimating the age of the mare surface into which Ranger 7 would crash. Hartmann counted the craters shown on photographs of the moon. To estimate the rate of crater production, he used the craters of the ancient Canadian Shield. Hartmann came up with an age for the mare surface of 3.6 billion years—as old as any terrestrial rock known at the time.

In order to choose the safest landing site, engineers needed more information than could be provided by a spacecraft crashing into the moon's surface, no matter how accurately they had aimed it. Practice landings by an unmanned craft could prepare the way for astronauts. A spacecraft that settled gently enough to survive the landing could telemeter scientific data about the surface back to Earth. Thus was born Surveyor.

Gold criticized Surveyor's design, claiming that when it landed, "even the antenna would sink out of sight." But Surveyor 1 landed in Oceanus Procellarum, bounced gently a few times, and then transmitted its first picture: an image showing the craft's footpad resting high and handsome in a small dimple. The next few Surveyors were equally successful and proved that the surface of the moon would bear a heavy weight. This vital question answered, mission planners could program Surveyors V and beyond for science.

The chemical analysis made by Surveyor V of the surface of Mare Tranquillitatis, only 25 kilometers from what in July 1969 would be Tranquillity Base, caused cold-mooner Urey to reflect unhappily, "It's basalt, isn't it?" He knew that basalt is not a primordial rock but forms when the interior of a body melts and differentiates. The presence of basalt showed that the interior of the moon had once been hot enough to melt. A few days later, Urey had recovered his equanimity and good humor, remarking, "Mother Nature knows best." John O'Keefe, the main "tektite from the moon person," as Urey referred to his many opponents on the source of these glassy travelers, was also disappointed, since tektites have a much different composition than basalt. O'Keefe responded to the Surveyor data by moving his putative tektite source from the maria to the lunar highlands.

The Surveyors and subsequent missions found that, over billions of years, myriad impacts large and small, had churned—or, gardened, in Shoemaker's phrase—the surface of the moon into a layer of debris. Even so, it appeared firm enough to support a spacecraft and its precious cargo. Thus, a manned landing appeared possible, but where on the moon to make it? The choice of the safest landing site required a much higher photographic resolution, and a more comprehensive view of the moon, than could be provided by Rangers crashing into it or Surveyors landing on it. "Surveyor had to land 'blind' . . . no one was willing to take a similar chance with Apollo," recalled lunar geologist Don Wilhelms.

The Lunar Orbiter mission would provide the needed eyesight by placing a spacecraft in continuous orbit around the moon and photographing the entire surface with a resolution, clarity, and coverage never before achieved even on Earth. The first three Orbiters succeeded in thoroughly photographing thirty-two possible Apollo landing sites, allowing the planners to program the next (and last) two Orbiters mainly for broad coverage and for science. Lunar Orbiters 4 and 5 captured almost the entire surface of the moon on film, in the process discovering the giant Orientale basin, a bull's-eye nearly 1,000 kilometers wide. Geologists could now see that impact had generated many of the unusual craters and other features they had ascribed to volcanism. The number of features attributable to volcanic activity shrank rapidly.

In the evening of July 20, 1969—prime time in the United States—Neil Armstrong stepped down from the Lander onto the surface of the moon. On its first try, our species had successfully left its birthplace and arrived at the surface of another body in the solar system. Armstrong had conceived a memorable quote but in the emotion of the moment did not deliver it well, an outcome familiar to any public speaker, and he was speaking to and for the entire human race. "That's one small step for a man, one giant leap for mankind," Armstrong intended to say, but he forgot to include the "a." No matter, let us insert it for this courageous pioneer who took us to the moon in an otherwise perfect mission.

When scientists opened the Apollo 11 sample box, they found, not the shimmering crystal of the Greeks, but what looked at first glance like dusty charcoal briquettes, and at second glance like terrestrial basalts. Microscopic and chemical analysis showed that the Apollo 11 rocks indeed were basalt, outwardly just like those of Earth. The radiometric age of the Tranquillity Base basalts came in at 3.6 billion years, the same as Bill Hartmann's estimate.

Though outwardly like terrestrial basalts, internally the moon rocks turned out to be surprisingly different. They were bone-dry, without a trace of water. Geologists had never seen such unaltered, crystal-clear minerals in their microscopes. In a rare moment of silliness, Urey had accepted the suggestion that the flat surfaces of the maria might be due to the accumulation of lake sediments. But the absence of even hydrated minerals in the lunar basalts showed that the moon had not only never had lakes—other than ice delivered by comets, the moon may have never had a single drop of water.

Compared to the earth, the moon has less of all the volatile chemical elements—those that boil off or evaporate easily—and more of the refractory ones. Apparently, at some time in its history, the moon melted and boiled off the volatile elements, leaving the refractory ones behind. The Tranquillity Base basalt was especially depleted in the rare earth element europium. The common mineral plagioclase feldspar takes up europium preferentially; somewhere on the moon, the geologists reasoned, must reside the large reservoir of plagioclase that had absorbed the missing europium. Since this reservoir did not seem to be in the maria, perhaps it lay in the highlands. Study of the lunar soil did turn up rare, plagioclase-rich particles, which had to come from somewhere.

Craterlets as small as a few microns (one millionth of a meter) pocked the surfaces of the lunar minerals. The difference in size between the largest impact basins like Orientale and these minuscule pits was over 1 trillion times, yet the same process, meteorite impact, created them all.

Origin of the Moon

The analysis of samples from subsequent Apollo missions confirmed the initial findings from the Apollo 11 rocks. The moon is ancient, at 4.5 billion years as old as the earth and the solar system. The mare basalts sampled and dated by Apollo appear to be among the youngest lunar features, yet all are older than all but the oldest terrestrial rocks. Neither tektites nor meteorites come from the moon (though a few specimens of a special class of meteorites from the moon have since been discovered on Earth). Debris gardened by impact, rather than a thick layer of dust, covers the moon. Impact created all lunar craters and almost every other feature seen and sampled there, save for the thin layer of basalt that fills the maria. Some regions of the interior of the moon were once hot enough to melt, but are now cold. The highlands are indeed made largely of plagioclase feldspar.

The presence of so much plagioclase provides a critical clue to the moon's origin. The only way for plagioclase to concentrate is to crystallize from a magma and then float to accumulate as a surface scum. But since the volume of plagioclase produced in this way is only a fraction of the volume of the magma from which it crystallized, a mountain of plagioclase requires an ocean of magma. The existence of the plagioclase highlands means that the moon must have melted to a depth of hundreds of kilometers, if not all the way to its center.

The europium provides another important clue. When no water or other source of oxygen is present, europium has a +2 ionic charge, the same as calcium. Since it then has about the same size as a calcium ion, europium substitutes for calcium in feldspar and other minerals. When oxygen is available, as it normally is, europium takes on a +3 charge and cannot substitute for a +2 element of a different size. The presence of the europium in lunar plagioclase, substituting for calcium, tells us that it existed in the +2 state, and therefore that the moon did not just lose water, it began without it. If ever water and other volatiles were present, they were lost in the process that created the moon.

The principal scientific justification for the Apollo program, which cost upward of $25 billion in the dollars of its day, was that by sending men to the moon and returning them and the samples they had collected to Earth, we would learn how the moon was born and thus, possibly, come to understand how our own planet formed. But in spite of all the new information and the implicit promise of the space program, Apollo failed to corroborate any of the three theories of lunar origin: capture, double-planet, or fission.

Capture remained nearly impossible to get right: either the moon would have escaped the earth's gravity and passed on by, or it would have collided with the earth. One important new discovery was that whereas the abundance of oxygen isotopes varies widely in meteorites, in rocks from the moon and the earth it is the same, suggesting that the moon and the earth formed near each other. Exploration of the capture hypothesis did lead to recognition that a wandering object caught by the earth's gravity would have been broken into pieces, which might then have aggregated into the moon.

The double-planet theory failed to explain the chemical differences between the earth and the moon, the rapid spin of the pair, and why the moon had melted. On the other hand, it did explain the oxygen isotope similarities.

At first, fission appeared the most promising of the three theories. Computer simulations showed that the rapidly spinning earth, instead of flinging off a blob of material, would have ejected a cluster of smaller fragments. These might have settled into orbit as a ring of debris, like that imagined long ago by Gilbert, which in turn coalesced into the moon. If fission had taken place after the earth's core had formed, the crust would have been depleted in iron. Then the moon, drawn from the earth's surface, would also be poor in iron, as we find it. Since the earth and the moon would have been made of the same material, they would have similar oxygen isotope compositions. So far, so good for fission, but as always, the familiar nemesis of angular momentum awaited. To fling off the ring of debris necessary to make the moon, the earth would have had to be spinning much more rapidly than the Earth-Moon system is spinning now, with no obvious way of getting rid of the excess spin once it was no longer needed.

The data from Apollo appeared to falsify each of the three classic theories and severely constrain any theory. Since the moon is undeniably there, staring us in the face, somehow it formed. Whatever has happened, can, and scientists would have to invent a new theory. A successful one would have to incorporate the new evidence and the best features of each of the classic theories, while avoiding their venerable flaws.

Perhaps the most fundamental fact about the moon is that it and the earth are members of the same solar system. Instead of trying to work out the origin of the moon directly, it made sense to develop a robust theory for the origin of the solar system, out of which the origin of the moon might fall as a byproduct.

15

A Most Difficult Birth

My mother groan'd, my father wept,
Into the dangerous world I leapt;
Helpless, naked, piping loud,
Like a fiend hid in a cloud.

—William Blake

The great students of the cosmos—the Greeks, Buffon, Kant, Laplace, Chamberlin, Moulton, Jeffreys, and others—all invented theories to explain the origin of the solar system. Some theories required another heavenly body, such as a passing star, to intervene. Others depended on our sun alone.

The nebular hypothesis of Kant and Laplace needed only a rotating sun, but it failed to explain why the sun has almost all the mass in the solar system, while the planets have almost all the angular momentum. If the sun had flung off the planets, either it should be spinning more rapidly or the planets should be revolving around the sun less rapidly, or both. This and other deficiencies sent scientists back to the old idea that another body had helped to create the solar system.

Buffon proposed that material dragged from the sun by a passing comet condensed into the planets. In the early twentieth century, Chamberlin and Moulton modified his hypothesis by proposing that the passerby was a star that pulled gases from both itself and the sun. As these gases cooled, they condensed into small, cold planetesimals that aggregated into the sun and planets. The passing-star theory failed when it was shown that material drawn from the sun would have had a temperature of over 1 million degrees and, instead of coagulating, would have flown off into space. Nevertheless, the Chamberlin-Moulton theory did get astronomers thinking about a solar system that began as a set of cold planetesimals. Perhaps those same starting conditions could derive in some other way than through the influence of a passing star.

By the 1950s, astronomers had observed large clouds of gas and dust—possible incubators of planets—in many regions of space. This made it seem that suns and planets need not be born by some random, improbable process, as when one star encounters another, but are the inevitable products of cosmic events. Solar systems might even be common. If this notion is correct, somehow the gas and dust must stick together to form larger masses, which collide and adhere to form still larger ones, and so on until finally are left planets, moons, and the scraps—the comets and asteroids—that escaped aggregation. Thinking about the final stages of this accretionary process put several scientists onto what in hindsight we can see was the right track.

Alfred Wegener and Robert Dietz show us that a scientist willing to challenge orthodoxy in one area (continental drift) is apt to do so in another (origin of lunar craters). In 1911, one year before Wegener published his first articles in German, Howard Bigelow Baker, a student of Schuchert's, published an article in the *Detroit Free Press* called "Origin of the Moon." In this and a series of papers, Baker conjectured that the moon had fissioned from the earth as a result of a close encounter with a now vanished planet. He imagined that the moon had been ripped from the Pacific Basin and that the resulting disturbance of the earth's crust caused the continents to drift.

More than four decades later, Baker published a pamphlet with the Detroit Academy of Sciences in which he concluded "that the moon was forcibly separated from the earth by some extraneous force is indicated by its excess angular momentum about the earth." The separating force might have come from a close encounter or a collision with some other heavenly body, which "bit deeply into the Pacific Hemisphere." Baker no longer connected fission with continental drift, but like his many predecessors, sought the source of the energy to drive earth dynamics.

No one seems to have paid any attention to Baker's theory, perhaps in part because he appealed to catastrophism, which always appears ad hoc and therefore suspect. To worsen his chances, Baker wrote at a time when catastrophism was at a low in popularity. In 1950, Immanuel Velikovsky had published *Worlds in Collision,* with its many fantastic and unsupported claims, such as that Venus had once passed so close that it grazed the earth. According to Velikovsky, this happened not millions of years ago, but within the last few thousand years—within recorded history. The scientific world cast Velikovsky out and ignored Baker.

Howard Baker had the germ of the right idea, but was ahead of his time. Almost twenty years later, another Baker, James, an engineer who designed cameras to track satellites, took Howard Baker's theory a giant step farther. James Baker proposed that a massive proto-Jupiter had disturbed the orbits of Mars and Earth, causing them to strike each other a glancing blow. Calculations by a colleague of James Baker's suggested that if such a grazing collision had happened, material dragged out of Mars would have settled into orbit around Earth, whence it could have condensed to form the moon. James Baker's idea illustrated one way in which a cloud of debris could be launched into orbit around the earth. Consideration of the failed fission theory had shown how such a cloud, once created, might aggregate into the moon. Pieces of the puzzle were beginning to fall into place.

James Baker presented his giant impact theory to astronomer Donald Menzel of Harvard University in 1973. Menzel was favorably impressed and reported Baker's ideas to a meeting of astronomers in Poland; later, he wrote that the theory "appears to combine the best features of all earlier hypotheses." Baker submitted a manuscript to *Reviews of Geophysics and Space Physics*, which rejected the article as "too speculative." The editor did suggest dividing the paper into smaller sections and publishing each independently, noting that "Your ideas have respectable company. Just this March John Wood of Harvard presented a tidal 'rip-off' hypothesis for the origin of the moon similar to yours in several respects." But most seem to have been less generous, viewing Baker's idea as "bordering on the crackpot." His theory remained unknown and apparently played no part in the eventual success of the identical theory.

Giant Impact

In 1974, the International Astronomical Union held a colloquium at Cornell University. At the time, William Hartmann was a youngster of thirty-five. Counting craters had made him aware that the moon, and presumably the other terrestrial planets, had suffered a massive bombardment by giant meteorites in the first few hundred million years of the solar system. Even the giant impactor that created the vast Orientale basin would not have been the largest of the bodies moving near the earth in the final stages of solar system formation. Where had these huge bodies gone?

Hartmann thought that the answer lay in the work of one of the few scientists who during the 1960s and 1970s was attempting to quantify the accreting planetesimal theory: Victor Sergeyevich Safronov, of the O. Yu. Shmidt United Institute of Earth Physics in Moscow. Working in isolation from Western scientists, Safronov invented sophisticated models for how the accretion of impacting planetesimals might have proceeded. The West learned of his work in 1972 with the translation of his book, *Evolution of the Protoplanetary Cloud.*

In the early 1970s, influenced by the work of Safronov, Hartmann and his collaborator Daniel Davis began to calculate the size distribution of the bodies in the last sweep-up of the solar system, just before the final set of planets and moons emerged. In each region, one large body (the future planet) would have been present, along with a smaller one, a still smaller one, and so on down to relatively tiny objects that had so far avoided accretion. The smallest of all, the original planetesimals, would remain and become the comets and asteroids. The second largest body traveling near the proto-earth would have been 500 to 3,000 kilometers in diameter; tens of bodies with radii larger than 100 kilometers would have been nearby. The diameter of our moon at its equator is about 3,500 kilometers; Mars measures about 6,800 kilometers; Earth about 12,500 kilometers. In other words, near the early earth, bodies the size of Mars and the moon flew though space at cosmic velocities. Immense collisions that would settle a planet's fate for the next 4.5 billion years awaited.

Hartmann and Davis realized that "the probability of the planet interacting with a large body is much larger than has been considered." They found that just half the energy of a body 1,200 kilometers in diameter, colliding with the earth at the expected speed (which they assumed was 13 km/sec, or 29,000 mph), was enough to launch two bodies the size of the moon.

When Hartmann rose to present his results at the 1974 Cornell conference, he speculated that if a body the size of Mars "hit the earth it might splash up the sort of material you need to make the moon and also get the right amount of kick into the system." The "kick" referred to the downfall of all other theories: the angular momentum of the earth and moon. The kind of glancing collision that Hartmann envisioned would have sent both bodies spinning like gargantuan tops, providing them with plenty of angular momentum.

Like the great Rutherford seventy years earlier, Hartmann feared the worst when he spotted in his audience a much more senior, highly respected authority. A. G. W. Cameron of Harvard is "a large, imposing man, who talks in a slow, pontifical fashion that might indeed strike fear into the heart of a young astronomer." Unlike Lord Kelvin at Rutherford's speech, Cameron was wide awake. But to Hartmann's relief, Cameron said, "We're working on the same idea, and we're coming to the same conclusions." Hartmann and Davis had been trying to work out the size distribution of objects just before the final sweep-up. In contrast, Cameron and his colleague William Ward had been trying to explain the angular momentum of the solar system starting from the collision of planetesimals. From different angles, the two pairs independently reached the same conclusion: violence on an unimaginable scale had marked the final stage of solar system assembly.

Cameron's scheme showed how the solid material blasted off the earth reached orbit, rather than, as might have been expected, falling back. The key is that the impact would have generated such temperatures—as much as 8,000 degrees Centigrade—as to vaporize the outer regions of the earth and the impactor. Cameron and Davis found that the rapidly expanding gases, jetting above the proto-earth, could have lofted more than twice the moon's mass into orbit. The effect resembles what happens when a satellite and its booster are launched atop a giant rocket and the booster is then ignited in a "second burn" to send the package on into orbit or space.

The giant impact theory did indeed combine the best features of its predecessors. As in the modern version of the capture theory, the moon accreted from fragments. The fragments came not from an alien object, but from the collision between a Mars-sized impactor and the proto-earth. As in the double-planet theory, prior to the collision, the impactor and the proto-earth were moving in the same region of space, having aggregated from the same part of the original dust cloud. As in the fission theory, the moon was born from the earth, though it had two parents, not one.

The moon's iron deficiency arose because, by the time of the collision, both the impactor and the earth had already formed iron-rich cores, leaving little iron at their surfaces, from where the moon was drawn. Because the moon bears the chemistry of both parents, it had been hard to tell whether the chemical similarities between the earth and moon outweighed the differences. The impact generated so much heat that the moon melted to its center, expelling volatiles and allowing a scum of plagioclase feldspar to soak up europium, float to the top of the magma ocean, and freeze to become the highlands.

As the New Zealand scientist and author S. Ross Taylor put it, the giant impact theory "cut the Gordian knot" that had fettered the three classic theories for the origin of the moon. Taylor referred to the legend of Alexander: When confronted with an intricate knot whose untying would identify the future conqueror of Asia, Alexander drew his sword and sliced it through. After centuries of thought had failed to solve the problem of the moon's origin, the giant impact theory did so in a single bold stroke.

Hartmann thought, naturally, that a theory that explained so much, neatly cleaving Gordian knots, would quickly earn a consensus. But he later decided that he had been naive: "No one had paid attention to our 1975 paper or Cameron's in 1976," he said. "I had thought that all you had to do was write a paper, and that was that—it would sink or swim on its own merits. But that's not so. You have to push a new idea."

Kona

Hartmann was not naive, only premature: behind the scenes, many were at work elucidating the giant impact theory. One of the most influential was George Wetherill, director of the Carnegie Institution of Washington, D.C. Wetherill had been inspired to think about giant impacts by photographs of the heavily cratered surface of Mercury sent back during the Mariner 10 mission in the mid-1970s. Mercury's battered face strengthened the case that the final stages of formation of the terrestrial planets was a battle of titans. When Wetherill learned of Safronov's ideas, he began to test them using his own computer models. His calculations, reported in 1976, confirmed those of Hartmann, Cameron, and their colleagues by showing that much of the mass of the early solar system would have been held in bodies about one tenth the size of the primary body: the protoplanet. Ironically, the United States beat the Russians to the moon, yet Safronov beat American scientists to the correct theory of its origin.

By 1984, fifteen years after Apollo 11, the time seemed ripe to convene a conference to discuss the new findings. Hartmann was one of the organizers. "When I went to the planning session for the conference to look over the abstracts for the proposed papers," Hartmann recalled, "I found, to my amazement and joy, that eight or ten of the abstracts—independently of each other—were about the impact idea." Almost to a person, the presenters offered new support from their own work for the giant impact theory.

Hartmann addressed one familiar criticism: the giant impact theory appeared ad hoc, ascribing the moon's origin to a one-off event dreamed up solely to explain it. Hartmann countered in a section of a paper entitled "Stochastic Does Not Equal Ad Hoc" ("stochastic" being a fancy word for random). "My point was," he said, "that although we cannot predict when or where a catastrophic event will happen, we can be sure that catastrophic impact events were happening all over the solar system, and therefore we shouldn't rule one out on the basis that it is ad hoc."

As the science writer Richard Kerr put it, the giant impact idea "breathed new life into a long-stagnant field." It relegated the classic theories of lunar origin to history. Nevertheless, several papers presented contrary ideas and evidence, and even the proponents of giant impact noted its shortcomings.

Summarizing in 1987, three years after the Kona conference, Cal Tech scientist David Stevenson gave a sober assessment: "A lot more work is needed. It is not yet clear whether the collision hypothesis satisfies the observational facts." Without naming their strategy, the new breed of scientists wisely adopted the old ideal of multiple working hypotheses.

Jeffrey Taylor, of the University of Hawaii, who along with Hartmann had helped to organize the Kona conference, summed up the results on his Web site:

> The conference was revolutionary. The traditional ideas for lunar origin were tossed aside by almost all attendees in favor of the giant-impact hypothesis. Most of us were dissatisfied with all of the old hypotheses. In [our] view, a giant impact was almost certain to happen. At the end of the three-day conference, the traditional hypotheses were discarded by most of us—a revolution in our thinking!

Monterey, 1998

One who has followed the ups, downs, and ultimate demise of the three classic theories for the origin of the moon, and the waxing and waning of theories for the origin of the solar system, could hardly be blamed for predicting that in time, the giant-impact theory would likewise accumulate flaws and collapse. Perhaps, as Schuchert claimed for the fit of the continents, the moon really is there to vex us and remind us of our human limitations. How ironic if the best minds could not explain the origin of the solar system's most viewed object.

To gauge the fate of the theory, we can pick up the story again at the next major conference on the origin of the moon, held at Monterey, California, in 1998. Several from the 1984 conference presented talks or posters, including Cameron, Hartmann, Stevenson, Taylor, and Wetherill, along with newcomers such as Robin Canup of the Southwest Research Institute in Boulder, Colorado. All had been pursuing the giant impact theory, taking advantage of the steady increases in computer power that allowed simulations to model more closely the complexity of the actual event. Far from meeting the ignominious fate of previous theories, by the end of the century, twenty-five years after it was first proposed, the giant impact theory had grown stronger.

Canup and Cameron found that at the time of the giant impact, the earth would have been only about half as large as today. The absorption of an impactor the size of Mars would have enlarged the earth to about two thirds of its present size; subsequent impacts would have built it up to the size we know. The computer simulations showed that the moon-generating impact took place in a bizarre double whammy. After the first terrible strike, some of the core of the impactor settles into the earth, but most bounces off, as if making a last-ditch effort to escape the clutches of the giant obstacle in its path. But escape from a body with the gravitational pull of even a half-earth proves impossible and the impactor falls to Earth again. The earth then melts completely, subsuming the iron core of the impactor. All this takes place in one-half hour!

The cloud of debris splashed out by the impact is lighter than the dense core of the impactor and, benefiting from the second burn, is lofted far enough away to avoid being swallowed up by the proto-earth.

In ten years or less, this debris coagulates into the moon, which melts completely. Just as we get used to a geologic time scale of billions of years, we learn that the most fundamental events of all took place on our human time scale. As accretion continues, the earth and moon attain their present size, cool, and find themselves nearly alone in their region of space, except for the leftover comets and asteroids that will continue to strike them without end. The moon's pristine rocks still preserve its early history; the earth's is lost forever, erased by the geologic activity of our living planet.

The giant impact theory has profound implications, not only for understanding how the planets and moons came to be, but for whether the seemingly random set of circumstances that produced them, and that, billions of years later, gave rise to *Homo sapiens,* could ever have been duplicated. Giant impact raises the ultimate question, as put by S. R. Taylor: Are we here by destiny, or by chance?

16

Impact Revolution

What is this talked-of mystery of birth
But being mounted bareback on the earth?

—ROBERT FROST

RECOGNITION OF THE importance of meteorite impact in the birth and life of the solar system constitutes a scientific revolution as profound as any in the earth and planetary sciences. Indeed, considering the import of the concept for how we see ourselves, our planet, and the likelihood of intelligence elsewhere in the universe, the impact revolution may be the most profound.

Science began the second half of the twentieth century denying meteorite impact a role in the origin of even the most obvious lunar and terrestrial craters. By century's end, scientists credited impact with the origin of the moon and of nearly everything on its surface. Most believed that impact caused one of the greatest mass extinctions in earth history, at the end of the Cretaceous period. The giant impact theory explains features of the solar system that have puzzled scientists for generations; indeed, the very configuration of the solar system as we know it is due to impact. When Gene Shoemaker said, "The impact of solid bodies is the most fundamental of all processes that have taken place on the terrestrial planets," he was more right than he could have appreciated. Without impact, almost nothing that is important in the solar system would be the same.

The inner planets and the asteroid belt alone contain enough enigmas to confound any theory. Mercury has no moon, no atmosphere, and the highest density of any body in the solar system. Venus also lacks a moon, but has a thick atmosphere. It rotates so slowly that a day is longer than our year. Earth has an atmosphere; its large moon is depleted in iron and once melted completely. Mars is just over half as large as the earth and has two small, insignificant moons. Beyond Mars, on the way to Jupiter, where we might expect to meet a planet, we find instead only a belt of asteroids. The giant impact theory can explain each of these puzzling and discrepant observations.

Part of the explanation of the differences is revealed by computer modeling, which shows that whereas a glancing blow between bodies of different sizes generates enough debris to form a moon, a head-on collision, or one between two objects of nearly the same size, does not. As we have seen, in the final sweep-up, a Mars-sized object struck the proto-earth a glancing blow and set both spinning rapidly, like tops given a sharp flick on their sides. The resulting cloud of debris, derived mainly from the iron-depleted outer regions of the proto-earth and its impactor, melted and coagulated into the moon.

Another giant impact blasted away most of Mercury's mantle and atmosphere, leaving it with an iron core, remnants of its original mantle, and no moon, crust, or clouds. The greater percentage of iron then left in Mercury raised its overall density. Thus, the same process that left Mercury rich in iron left our moon depleted; everything depended on just when and how the last few large objects in the primordial solar system were swept up.

Venus may have been struck head-on and stopped dead in its

tracks, or it may have escaped being hit, leaving it with only the slow spin of the primordial dust cloud. The incessant rain of volatile-rich comets throughout geologic time may have delivered the atmospheres of Venus and Earth.

Until the last few decades, students of the solar system have assumed that the asteroid belt, from which our meteorites come, represents an exploded planet. Today, we know that the sum total of all the mass of the asteroids amounts to only about 5 percent of the mass of our moon. So small a body could never have differentiated into a true planet. But why did a planet not form at that location? The answer is: Jupiter.

Growing up near Jupiter was like living next door to a voracious bully who stole the morning milk delivery from your doorstep and robbed your pantry late at night, stunting you and your family. Jupiter was born at just the right distance from the sun to grow, like the Ugly Duckling, too large, too fast. The faster it grew, the more gravitational pull it exerted on nearby objects, and the faster still it grew. Any planetesimals that began to aggregate between Mars and the young giant found themselves first ripped to pieces by Jupiter's massive gravity, then sucked into its greedy maw, to disappear forever. So little material escaped Jupiter that Mars became a dwarf. Jupiter demanded sacrifice.

Without a Jupiter, or with a smaller one, a planet would exist where now we find the asteroid belt. Mars would have absorbed many more planetesimals and might have grown as large as the earth. Instead of Earth, with its moon, and Mars, with its two small moons, we might find two medium-sized planets, neither with a moon.

But sometimes, as a bully grows up, he matures into a protective big brother. Jupiter has warded off from Earth a succession of potential life destroyers. This became ever so clear in July 1994, when the "string-of-pearl" fragments of comet Shoemaker-Levy 9 crashed into Jupiter one after another. Just one of the impact fireballs would have engulfed the entire earth.

The Mars-sized object that struck the earth tilted our planet's axis and sent it spinning at a rate that eventually settled to twenty-four hours per revolution. Had the impact not occurred, or had the axis been more nearly vertical, Earth would have no seasons. Then temperature and climate would vary only with distance from the equator. How this would have affected our evolution and culture no one knows. What we do know is that the twenty-four-hour diurnal cycle is deeply embedded; even when placed in a windowless room for weeks, we retain its imprint. The menstrual cycle of human females, 29.5 days, records some mysterious obligation to the moon, whose complete cycle is also 29.5 days. With some other moon, or, in the words of the song, "no moon at all"; with some other axial tilt, or no tilt at all, the course of earth history and evolution would have proceeded in some other, unknowable way.

By counting crater densities and measuring the ages of returned samples from the moon, we know that from the birth of the solar system 4.55 billion years ago until about 3.8 billion years ago, the bombardment of the four inner planets was especially intense. Even in the last 100 to 200 million years of this Hadean eon, there formed on the moon at least 1,700 craters greater than 20 kilometers in diameter, and at least twelve basins larger than the Yucatan Chicxulub crater, which marks the impact of the dinosaur killer. Even more meteorites would have struck the larger target that the earth provided.

It may be no accident that the oldest remains of life also date to 3.8 billion years. As pointed out in Peter D. Ward and Donald Brownlee's fine book, *Rare Earth* (2000), impact may have thwarted the development of advanced life, while at the same time providing a haven to protect primitive life until the intense early bombardment ceased. Within the last few decades, scientists have discovered that life can flourish even in extreme terrestrial environments. In the deep-sea smokers that pop up on the mid-oceanic ridges, in the steaming pools of Yellowstone, thermophile bacteria thrive at temperatures of hundreds of degrees. Some believe that these heat-lovers were the first complex life forms. A large impact, especially of a water-rich comet, fractures, heats, and lubricates the rock at ground zero down to a depth of several miles. These piles of broken, wet rock may stay hot for a million years, serving to incubate thermophiles.

By bringing the necessary organic compounds to Earth from space, impact may have seeded our planet with life. Instead, life may have begun somewhere else—on Mars, for example, and traveled to Earth. We know that impact has blasted pieces off Mars and the moon and sent them ricocheting to Earth. Our planet surely sent its own impact-generated meteorites to Mars and the moon. Scientists have claimed that one Martian meteorite found on Antarctica contains fossil bacteria; recent studies show that Martian bacteria could have survived the trip. Furthermore, detailed calculations show that more than 5 billion Martian rocks capable of bringing life fell on Earth in the last 4 billion years. The researchers concluded that "if microbes existed or exist on Mars, viable transfer to Earth is not only possible but also highly probable. Earth-to-Mars transfer is also possible but at a much lower frequency." Impact may have colonized the solar system. Perhaps only on our blue planet could the colony survive for long, and evolve.

Impact may or may not giveth, but it certainly taketh away. Sixty-five million years ago, a meteorite at least 10 kilometers in diameter, traveling at 15–30 kilometers per second and therefore carrying the energy of a billion Hiroshima bombs, struck the Yucatan peninsula. Even though a kilometer of rock now covers the crater, seismic or gravity measurements allow us to visualize the resulting crater as plainly as we can see our arm bone in an X ray. Exactly simultaneously with the impact, 70 percent of species died.

One can attribute the simultaneity of the impact and the K-T extinction to cause and effect, or one can claim that it is coincidence, as a few scientists continue to do. If one chooses to accept the considerable evidence for cause and effect, then impact brought an end to the life span of a vast number of living creatures. At the moment the dinosaur killer struck, genes and fitness no longer mattered. The dice had been rolled, and most living creatures would lose. But not all lost: a few who were in the right place at the right time, or whose previously marginal modus operandi would now prove more successful, held on.

The end-Cretaceous extinction was one of five great mass extinctions in earth history. None of the other four shows convincing evidence of impact, though several show some evidence. The largest extinction of all ended the Permian period of the Paleozoic era, when some 90 percent of marine species died. Geologists have attributed the end-Permian extinction to a variety of causes, or to several acting in concert. Some believe that volcanic gases caused the extinction; a major basalt eruption did occur at the same time in Siberia. On the other hand, recent evidence suggests that the end-Permian extinction happened in only tens of thousands of years. No terrestrial process that we know of acts that quickly, though scientists may yet discover one that does. The release of deadly methane gas from the seafloor is one new possibility that they are studying.

Beyond its possible role in the largest mass extinctions, we can wonder what the effect has been of the thousands of lesser but still energetic impacts on Earth. In the last 300 to 500 million years, impact has produced 950 craters on Venus, one about every half-million years. The largest is Mead, 269 kilometers in diameter. Estimates based on astronomical observations and on terrestrial craters suggest that every 2 to 3 million years, a meteorite large enough to create a 20-kilometer crater strikes the earth. At that rate, more than two hundred such craters would have formed since the beginning of the Paleozoic. Larger impactors would do more damage but arrive less frequently. To say that hundreds or thousands of large meteorite impacts had no effect on the evolution of life seems counterintuitive, though intuition has led us astray before.

Scientists took so long to recognize the importance of impact because it occurs on a completely different time scale than our own and because the evidence is so quickly removed. But now we have enough facts to allow a thought experiment that will remind us of the frequency and importance of impact in earth history.

Imagine that a video camera mounted in space had photographed the earth during each moment of its 4.5-billion-year span, and that we can play back the entire history in one hour. Viewing the vastly speeded-up video, we would see the Hadean eon close and the magma ocean cool. Elephantine continents would emerge and dance clumsily over the surface, colliding to throw up mountain ranges and splitting apart to form ocean basins, all the while accreting steadily along their margins. But something else would be evident that has left little trace: the steady rain of comets and asteroids into the earth, each throwing up a blaze of light that hardly has time to dim before the next flashes. Most of the impactors would strike in the ocean basins and disappear; all but the largest that landed on continents would have their craters worn away by erosion, or covered by sediments, in a few seconds. If we happened to see, out of the corner of our eye, the crater Tycho form on the moon 100 million years ago, a moment later we would witness a large blast on Earth. It would be all too visible, for it blazed with the light of a million suns. This impact would cause a wrenching change in the direction of evolution, allowing a group of small, hamster-sized creatures to supersede the giant reptiles that had dominated for the previous few seconds of hypertime. An instant before the tape ran out, too fast for us to glimpse, the descendant of these lucky mammals would leave the trees, stand erect, make tools, and speak.

The incessant flashes of impact-generated light would be apt to be the most memorable feature of the video. They would remind us that the earth is not isolated in space, but is part of a planetary system in which impact is ubiquitous, incessant, and fundamental. We would see the present for what it is: merely the last moment out of innumerable moments in the history of our planet and our solar system. The viewer would know, having seen it, that an entire process has gone unnoticed because of our allegiance to that fleeting, uncapturable moment called the present.

During the space missions of the 1960s, for the first time we could witness our small blue-white planet, floating alone and lovely in space. In the photographs, the earth appears isolated and unaffected by any other bodies—an entirely self-contained world. If such it were, the present would more nearly be the key to the past and geologists would not need to look to the heavens for the causes of earthly events.

But the earth appears isolated, unaffected by objects from space, only because of our circumscribed time horizon. We define the present on our time scale, in which a human life lasts a few decades, written history goes back only a few thousand years, and *Homo* only a few million. The founders of geology, not knowing that the earth is billions of years old, striving to overthrow biblical catastrophism, had nothing but the present thus defined with which to work.

To these pioneers, the recognition that the present is the key to the past went a long way toward converting their field into a science. But in the intervening two hundred years, their reasonable slogan became twisted. Instead of remaining a useful guide to the processes that shape the earth, it became proscriptive: Anything that we cannot observe cannot have happened. Continents cannot drift, meteorites cannot strike the earth and change evolution. But that is illogical and wrong. One cannot simultaneously accept that the earth is billions of years old and hold that every important process has taken place in the eyeblink of geologic time during which scientists happen to have been observing.

The earth was born in violence and has lived in a solar system marked by violence. Locked in the present as we are, when we look out, we fail to see the evidence. Indeed, the most dramatic effect we are likely to see is the birth and sudden death of an insignificant shooting star, blazing its way to cosmic oblivion. The moon passes through her predictable phases, the very symbol of constancy, her face always obediently turned toward us. The motion of the planets and their moons defines regularity, so reliably that, before global positioning systems, sailors used the positions of the planets and their moons to navigate. Out there, everything appears calm, orderly, and unchanging. So it might have looked to the dinosaurs, had they developed intelligence in their 160-million-year reign. Long experience would have justified a saurian complacency. But impact reigns—just give it time. Even billions of years in the future, impact will continue. Long after our sun has died and Earth's internal fires have been extinguished, the energy of impact will provide the only heat and light on our cold, dead planet.

17

The Tapestry

The moving finger writes, and having writ,
Moves on; nor all your Piety nor Wit
Shall lure it back to cancel half a Line,
Nor all your Tears wash out a Word of it.

—OMAR KHAYYÁM

OVER SEVERAL HUNDRED YEARS, starting with only the raw wool of thought, observation, and experiment, science has crafted many threads and woven them into a beautiful tapestry of knowledge. The earliest threads were observations; repeated confirmation converted them into the strongest threads of all: facts. The fibers that represent certain facts are as strong as steel: the speed of light, the gravitational constant, the frequency of harmonic oscillation of a quartz crystal, the rate of decay of uranium 238.

As facts accumulated, scientists invented theories to explain them and wove the theories into the fabric. Though facts, once validated, are a permanent part of the weave, theories are ephemeral, always at risk of replacement. As new observations and interpretations weaken a theory and remove it from the weave, its replacement arrives right behind it, strengthening the whole.

In spite of the evident human failings of scientists, the tapestry of science grows ever stronger. Egotism, prejudice, poor guesses, and outright mistakes, all in the end make no difference. Their effect is only to delay truth, not to deny it. The invisible moving finger of science weaves a tapestry far stronger than scientists themselves. Scientists and their theories come and go; the tapestry of science is eternal.

The thread that represents the age of the earth is one of the strongest, held firmly in place by crossing threads from astronomy, biology, chemistry, palentology, and physics. But the other two revolutionary discoveries of twentieth-century earth science also bind the thread of time. Without geologic time, geologists would have to abandon both plate tectonics and meteorite impact.

Many lines of research combine to show that the giant plates that segment the earth's exterior move at a rate of a few centimeters per year—about the rate at which our fingernails grow. The paleomagnetic time scale derived from volcanic rocks on land; the symmetrical seafloor magnetic anomaly stripes; the paleomagnetic record in deep-sea cores; the age of the deepest sedimentary rocks in different regions of the ocean basins: all crisscross perfectly as warp and woof. The simplest of calculations gives the Atlantic spreading rate. Divide the distance separating the continents on either side (about 6,000 km) by the age of the rocks at the margins (about 180 million years) and derive the rate: 3.3 centimeters per year. Using global positioning systems, laser ranging, and other techniques, scientists today actually measure the otherwise imperceptible motion of the plates. They obtain rates exactly the same as those they deduce from geology and geophysics. Plate tectonics unifies geology, but it needs time.

Throughout the thousands of years of recorded human history, no one has seen a meteorite fall and create a crater. Over the hundreds of years since Galileo, in spite of countless hours spent observing the moon through a telescope, no one has ever seen a new crater form there (though a few have thought they did). On Earth, impact craters are so rare that it took well over a century for geologists to recognize them. Evidently, craters form on a much longer time scale than our human one. Yet myriad craters from the immense down to the submicroscopic pock the moon's surface; even more mark the hidden far side. Craters cover the surfaces of Mars, Venus, and Mercury—indeed, they blanket every solid body that we have observed since the space age began, save those that are still volcanically active. A process so slow that we can never observe it happening, yet that has had such ubiquitous effects, must have been going on for a long time. To saturate an entire solar system with craters takes time.

Threads from all fields of science bind the tapestry into a tight and interdependent cloth. A single, strong, time-tested thread, especially one representing the fundamental laws of mathematics, chemistry, and physics, cannot be removed from the fabric, lest it pull with it many other threads, eventually shredding the entire weave. Extracting the thread that represents the age of the earth would pull out not only the threads of plate tectonics and meteorite impact, but those representing the essential facts and theories of the physical and biological sciences. For the tapestry to remain whole, the thread that stands for the age of the earth must remain.

The tapestry of science crowns our species. To tamper with it is no different from tampering with another great tapestry of human invention: the arts. Imagine removing every fourth note from *Eine kleine Nachtmusik,* every fourth line from *Hamlet,* every fourth brush stroke from Van Gogh's *Irises.* To do so would be to deny and desecrate the finest that our species has achieved.

Notes

Chapter 1

3 “**If an elderly but distinguished”:** Arthur C. Clarke, *Time.*

4 “**I came into the room”:** Arthur Stewart Eve, Rutherford; *Being the Life and Letters of the Rt. Hon. Lord Rutherford, O.M.* (New York: The Macmillan Company, 1939), 107.

5 “**I know,” said Rutherford:** Ibid., 23.

6 **Some said the stones:** Claude C. Albritton, *The Abyss of Time. Changing Conceptions of the Earth's Antiquity After the Sixteenth Century* (San Francisco: Freeman, Cooper, 1980), 24.

7 **Sensing the charm:** Ibid.

7 “**At the time when any given stratum**”: Ibid., 38.

7 **“Beautiful is that**”: Ibid.

9 **Then the priest:** Ibid., 73.

10 **Even that proved too much:** Stephen G. Brush, *Transmuted Past: The Age of the Earth and the Evolution of the Elements from Lyell to Patterson* (New York: Cambridge University Press, 1996), 4.

12 **“a great mistake has been made”:** Joe D. Burchfield, *Lord Kelvin and the Age of the Earth* (Chicago: University of Chicago Press, 1990), 81.

13 “**doth bestride my world**”: Stephen Jay Gould, *Time's Arrow, Time's Cycle: Myth and Metaphor in the Discovery of Geological Time* (Cambridge, MA: Harvard University Press, 1987), 179.

14 **Fourier calculated:** Brush, *Transmuted Past:*, 27.

14 **At age seventeen:** Ibid., 34.

14 "**Within a finite period**": Burchfield, L*ord Kelvin and the Age of the Earth,* 23.

15 "**It seems, therefore**": Ibid., 31.

15 "**Apparently Kelvin was infected**": Brush, *Transmuted Past*, 37.

16 "**By 1868**": Burchfield, *Lord Kelvin and the Age of the Earth*, 43.

16 **But like Kelvin and others:** Brush, *Transmuted Past*, 5.

16 **As an example of geological:** Ibid., 38.

17 **Lyell's Principles had so impressed:** Charles Darwin, *On the Origin of Species by Means of Natural Selection* (Cambridge, MA: Harvard University Press, 1964 facsimile edition), 282,

17 "**for a cliff**" . . . "**At this rate**": Burchfield, *Lord Kelvin and the Age of the Earth,* 71.

17 "**those confounded millions**": Ibid.

17 **Thomas Huxley:** Ibid., 1.

18 **Now Lyell even admitted:** Ibid., 68.

18 "**Catastrophes may be**": Ibid., 83.

18 "**Mathematics may be compared**": Ibid., 84.

19 **But after laboriously calculating:** Ibid., 96.

19 **Though not many occupied:** Ibid., 95.

20 **When the adjustments were over:** G. Brent Dalrymple, *The Age of the Earth* (Stanford, CA: Stanford University Press, 1991), 68.

20 **Samuel Haughton:** See Burchfield, *Lord Kelvin and the Age of the Earth,* 100–03.

21 "**The actual period**": Dalrymple, *The Age of the Earth*, 51.

21 "**Utterly impossible**": Burchfield, *Lord Kelvin and the Age of the Earth,* 109.

22 "**suffices to sweep away**": Brush, *Transmuted Past*, 37.

22 "**We have been drawing**": Ibid., 29.

23 "**It is difficult**": Ibid.

23 "**The utmost any physicist**": Burchfield, *Lord Kelvin and the Age of the Earth*, 126.

23 **As Kelvin relentlessly:** Ibid., 135.

24 "**Is present knowledge**": Ibid., 144.

25 **If one knew the rate:** Dalrymple, *The Age of the Earth*, 52.
25 **By the end of the nineteenth century:** Ibid.
26 **When we repeat:** Ibid., 58.

Chapter 2

29 "**An important scientific innovation**": Max Planck, *Scientific Autobiography* (New York: Philosophical Library, 1949), 00.
33 **In 1903, Rutherford found:** Eve, *Rutherford*, 89.
34 "**The discovery of the radio-active**": Burchfield, *Lord Kelvin and the Age of the Earth*, 164.
34 **At the subsequent meeting:** Eve, *Rutherford*, 109.
35 **In a series of letters:** Ibid., 141.
35 **Kelvin claimed that radium:** Ibid., 140.
35 "**Go on forever**": Burchfield, *Lord Kelvin and the Age of the Earth*, 165.
35 **Lord Kelvin died:** Eve, Rutherford, 161.
36 **Considering the possible errors:** Dalrymple, *The Age of the Earth*, 72.
36 "**Rutherford's achievements**": Eve, *Rutherford*, 430.
37 **Strutt measured the amounts:** Arthur Holmes, *The Age of the Earth* (London & New York: Harper & Brothers, 1913), 154–55.
37 **Though he found:** Dalrymple, *The Age of the Earth*, 72.
40 **Like Lord Kelvin:** Burchfield, *Lord Kelvin and the Age of the Earth*, 189.
40 "**Wherever the geological evidence**": Dalrymple, *The Age of the Earth*, 74.
41 "**With these discoveries**": Arthur Holmes, *The Age of the Earth, an Introduction to Geological Ideas* (New York: Harper & Brothers, 1927), 18.
41 "**From the mists**": Ibid., 166.
42 "**The discordance between**": Ibid., 170.
42 **Since "the modern hour-glass**": Ibid., 174.
42 "**incapable of providing**": Ibid., 44.
42 "**No more definite**": Arthur Holmes, "Radioactivity and Geological Time," Bull. Nat. Res. Council (Washington, DC: National Research Council of the National Academy of Science, 1931), 454.

Chapter 3

45 "**The time has gone by**": A. S. Eddington, *Nature*, 111 (1923).

53 **Nevertheless, on February 20:** E. K Gerling, "Age of the Earth According to Radioactivity Data," *Doklady Akademii Nauk SSSR* 34, no. 9 (1942).

54 **Using a complex approach:** Arthur Holmes, "A Revised Estimate of the Age of the Earth," *Nature*, 159 (1947).

54 **The third in the trio:** Iosif Kriplovich, "The Eventful Life of Fritz Houtermans," *Physics Today* (July 1992).

55 "**Hurry up**": Ibid., 35.

56 **Brown, inept to the point:** See Shirley Cohen, "Duck Soup and Lead," *Engineering and Science*, 1997.

57 **After graduating:** Ibid., 21.

57 **Like many others:** See C. Davidson, ed., *Clean Hands: Clair Patterson's Crusade Against Environmental Lead Contamination* (Commack, NY: Nova Science Publishers, 1999).

58 **Before the year was out:** F. G. Houtermans, "Determination of the Age of the Earth from the Isotopic Composition of Meteoritic Lead," *Nuovo Cimento*, 10, no. 12 (1953).

59 **Patterson reported:** C. C. Patterson, "The Isotopic Composition of Meteoritic, Basaltic, and Oceanic Leads, and the Age of the Earth." Paper presented at the Nuclear Processes in Geologic Settings, Williams Bay, Wisconsin, 1953.

61 **Rather than admit to this:** Cohen, "Duck Soup and Lead," 30.

Chapter 4

61 "**When you cannot measure**"**:** Lord Kelvin, "Lecture to the Institution of Civil Engineers," May 3, 1883.

62 **Upon hearing of the account:** Dalrymple, *The Age of the Earth*, 257.

64 **By the early 1990:** Ibid.

68 **As an example:** Ibid.

69 "a riddle wrapped": Winston Churchill, Radio broadcast (October 1, 1939).

70 **A group of over fifty zircons:** Dalrymple, *The Age of the Earth*, 188

70 **Further zircon studies**: S. A. Bowring, and I. S. Williams, "Priscoan (4.00–4.03 Ga) Orthogneisses from Northwestern Canada," *Contrib. Mineral. Petrol*, 134 (1999).

71 **The Isua, Greenland, "supracrustals":** Dalrymple, *The Age of the Earth*, 141.

73 **Yet because the volume:** Ibid.

Chapter 5

77 **"A science that hesitates"**: Alfred N. Whitehead, *Science and the Modern World* (New York: Macmillan, 1926).

77 **"If we are to believe"**: Van Waterschoot van der Gracht, ed., *Theory of Continental Drift*, 87

79 **"Seuss has secured":** Naomi Oreskes, *The Rejection of Continental Drift: Theory and Method in American Earth Science* (New York: Oxford University Press, 1999), 12.

81 **Dana argued that:** Ibid., 16.

81 **One of the most influential:** Ibid., 48.

81 "**The great ocean basius**": Ibid., 17.

82 **As Dana put it:** Ibid., 63.

82 "**The best geological ideas**": Ibid.

Chapter 6

83 **Chapter title:** P. Termier, "The Drifting of Continents," Ann. Rept. Smithsonian Inst (1925).

83 "**If at first**": A. Einstein (Amazon.com Free Coffee Mug).

84 "**The first concept**": A. Wegener, *The Origin of Continents and Oceans*. 4th ed. (New York: Dover, 1966) 1.

84 "**Wegener was a very**": Henry W. Menard, *The Ocean of Truth: A Personal History of Global Tectonics* (Princeton, NJ: 19

85 "**It is just as if**" : *Ibid.*.

86 **And then the seemingly:** Wegener, T*he Origin of Continents and Oceans*,

86 **Alexander du Toit:** See H. E. LeGrand, *Drifting Continents and Shifting Theories: The Modern Revolution in Geology and Scientific Change* (Cambridge & New York: Cambridge University Press, 1988).

88 "**We shall refrain**": p. 133

88 "**Newton theory of drift**"**:** Wegener, *The Origin of Continents and Oceans*, 167.

Chapter 7

91 "**To be uncertain**": Cited by Arthur Holmes, "Response to Penrose Medal Citation," *Proc. Geol. Soc. Amer.* 1956 (1957).

91 "**He is not seeking**": LeGrand, *Drifting Continents and Shifting Theories*, 62.

91 "**Science has developed**": Ibid.

92 "**After considering**": Bailey Willis, "Continental Drift," in *Theory of Continental Drift,* ed. W. A. J. M. Van Waterschoot van der Gracht 76, Tulsa, OK: American Association of Petroleum Geologists London: T. Murby & Co., 1928),

92 "**Can we call geology**": Van Waterschoot van der Gracht, ed., *Theory of Continental Drift*, 83–87.

92 **E. W. Berry . . . asked:**

93 **"auto-intoxication":** Oreskes, *The Rejection of Continental Drift*, 126.

93 "**He generalizes too easily**": Charles Schuchert, "The Hypothesis of Continental Displacement," in Van Waterschoot van der Gracht, ed., *Theory of Continental Drift*, 139.

93 "**taken extraordinary liberties**": Ibid., 111.

94 "**Are we to believe**": Ibid., 116.

94 "**It can be truthfully" and the quotes that follow:** Ibid., 117.

94 **Schuchert had not:** Oreskes, *The Rejection of Continental Drift*, 99.

94 **Daly was one of the few:** Ibid.

94 **In the biographical memoir:** Ibid., 89.

95 "**We are on safe ground**": Schuchert, "The Hypothesis of Continental Displacement," 117.

96 "**The South Atlantic land bridge**": Oreskes, *The Rejection of Continental Drift*, 193.

96 "**In spite of all the**": Ibid., 198.

97 **The pair of papers:** Ibid., 218.

97 **Willis wrote a paper:** Bailey Willis, "Continental Drift, ein Märchen," *American Journal of Science*, 242, no. 9 (1944).

Chapter 8

99 "**Anything that has happened**": Oreskes, *The Rejection of Continental Drift,* 63.

100 **The drastic horizontal shortening:** A. Hallam, *A Revolution in the Earth Sciences: From Continental Drift to Plate Tectonics* (Oxford: Clarendon Press, 1973), 151.

100 **People accepted electricity:** Ibid., 110.

102 **Joly thought that:** Oreskes, *The Rejection of Continental Drift*, 110.

102 **In his last major publication:** See John Joly, *The Surface-History of the Earth* (Oxford: Clarendon Press, 1930).

103 **In a talk:** Arthur Holmes, "Radioactivity and Earth Movements," Trans. Geol. Soc. Glasgow, 18 (1929).

104 "**Mountain building" and the quotes that follow:** Arthur Holmes, "A Review of the Continental Drift Hypothesis.," *Mining Magazine*, 40 (1929), 346.

105 **In 1927, Schuchert wrote:** Oreskes, *The Rejection of Continental Drift,* 193

106 "**It is along the lines**": Ibid., 198.

106 "**Pray don't get**": Ibid., 213.

107 **But Willis could not get by:** Willis, "Continental Drift, ein Märchen."

107 "**Fellow scientists**": Ibid.

108 **In his 1944 text:** See Arthur Holmes, *Principles of Physical Geology* (London & New York: T. Nelson & Sons, 1944), 487–509.

110 "**all the continents except**": Ibid., 503.

110 **The only argument:** Ibid., 504.

110 **Holmes ends his book:** Ibid., 508.

Chapter 9

113 "**The sea, washing**": Ralph Waldo Emerson, "Wealth," in *The Conduct of Life* (1860).

114 **They might be witnessing:** Oreskes, *The Rejection of Continental Drift,* 249.

115 "**Diametrically opposed**": Ibid., 254.

115 **He described a curiosity:** Ibid., 257.

115 "**Many delegates turned**": Ibid., 262.

116 **As the historian of science:** John A. Stewart, *Drifting Continents and Colliding Paradigms: Perspectives on the Geoscience Revolution* (Bloomington: Indiana University Press, 1990), 47–48.

118 "Wegener's alleged fit": Harold Jeffreys, Letter, *Nature*, 195 (1962).

118 **Cambridge University legend:** F. J. Vine, "The Continental Drift Debate," *Nature*, 266 (1977).

Chapter 10

123 "**How extremely stupid**": Quoted in K. Lorenz, *On Aggression* (London: University Paperback, 1967), 237.

125 **"was destined to become":** William Glen, *The Road to Jaramillo: Critical Years of the Revolution in Earth Science* (Stanford, CA.: Stanford University Press, 1982), 263.

128 **Matthews returned to Cambridge:** Glen, *The Road to Jaramillo*, 274.

128 "**We had no idea**": Ibid.

128 **Robert S. Dietz wrote:** Robert S. Dietz, "Continent and Ocean Basin Evolution by Spreading of the Sea Floor," *Nature*, 190 (1961).

129 "**chance favors**": L. Pasteur, "Inaugural Lecture" University of Lille, 1854.

130 **Vine conjectured:** Glen, *The Road to Jaramillo*, 278.

130 **Graduate student Vine:** Ibid., 279.

130 "**Virtually a corollary**" . . . "**If the main**": F. J. Matthews and D.H. Vine "Magnetic Anomalies Over Oceanic Ridges," *Nature*, 199 (1963).

131 **"like a lead balloon":** Glen, *The Road to Jaramillo*, 281.

131 "**Probably not adequate**": Ibid., 304.

131 "**Thought Vine and Matthews'**": Ibid., 279.

131 "**You don't believe**": E. Bullard, "The Emergence of Plate Tectonics; a Personal View," *Ann. Rev. Earth & Planetary Sciences*, 3 (1975), 20.

131 **Lawrence Morley was chief:** Glen, *The Road to Jaramillo*, 298

131 "**A natural thing**": Ibid., 301.

131 "**Such speculation**": Ibid., 299.

132 "**Probably the most significant**": Ibid., 302.

132 **The question of who:** Ibid.

133 **To fill this critical time:** Ibid., 262–66.

133 **Fred Vine was present:** Ibid., 310.

134 **"Ewing's philosophy"**: Ibid., 313.

134 **"The gods visit"**: Euripides, "Phrixus, Fragment 830."

135 **"Ewing remained"**: Robert S. Dietz, "Earth, Sea, and Sky; Life and Times of a Journeyman Geologist," *Ann. Rev. Earth & Planetary Sciences*, 22 (1994).

135 **Although the time was late . . . magnetic data as well:** Glen, *The Road to Jaramillo*, 332.

135 **They were at sea:** Ibid.

135 **There followed further ridicule:** Ibid., 334.

135 **"Well, that knocks"**: Ibid., 335.

136 **" It's so good"**: Ibid., 339.

136 **"a starving man"**: Ibid., 336.

136 **Unconvinced until that moment:** Ibid., 358, 336.

136 **"Well, God, he"**: Ibid., 337.

136 **"the most exciting year"**: Ibid., 339.

137 **"The entire history"**: Ibid., 340.

137 **In a hotel room:** Ibid.

137 **Undaunted, and with access:** Ibid., 348.

138 **"Three different features"**: Ibid., 351.

139 **In 1968, Tuzo Wilson:** Stewart, *Drifting Continents and Colliding Paradigms*, 111.

139 **John McPhee has captured:** See John A. McPhee, *Annals of the Former World* (New York: Farrar, Straus & Giroux, 1998).

141 **To his credit:** V. V. Belousov, "An Open Letter to J. T. Wilson," *Geotimes*, 13 (1968); J. Tuzo Wilson, "A Revolution in Earth Science," *Geotimes*, 13, no. 10 (1968).

142 **He and his co-author:** Charles B. Officer, et al., "Cretaceous-Tertiary Events and the Caribbean Caper," *GSA Today*, 2 (1992).

Chapter 11

145 **"Holmes was an anachronism"**: Quoted in Ursula B. Marvin, *Continental Drift: The Evolution of a Concept* (Washington, DC: Smithsonian Institution Press, 1973), 103.

145 **"The test of a first-rate"**: F. Scott Fitzgerald, "The Crack-Up," *Esquire* (1936).

146 "**Brazil cracked away**" . . . **We can explain**": Oreskes, *The Rejection of Continental Drift*, 196.

146 **In 1929, he wrote . . . and the quotes that follow:** Ibid., 198.

147 "**To end on a personal note**" . . . "While so many contradictory": Arthur Holmes, "The South Atlantic: Land Bridges or Continental Drift," *Nature*, 171 (1953).

147 **Dunbar:** is a professor of geology at Yale.

148 **Holmes cited two accomplishments:** Hedberg, "Penrose Medal Citation for Arthur Holmes," *Proc. Geol. Soc. Amer.* 1956 (1957).

148 **For whatever reason:** Menard, *The Ocean of Truth*, 87

149 "**The Russian geologist**": Arthur Holmes, *Principles of Physical Geology* (New York: Ronald Press Co., 1965), 1203.

149 **His biographer for the Royal Society:** K. C. Dunham, "Arthur Holmes," *Biographical Memoirs of Fellows of Royal Society*, 12 (1996), 294.

151 "**As a paleontologist**": Oreskes, The Rejection of Continental Drift.

152 **The same response came:** See James Lawrence Powell, *Night Comes to the Cretaceous: Dinosaur Extinction and the Transformation of Modern Geology* (New York: W. H. Freeman, 1998).

152 "**In the midst of this intellectual**": Claude J. Allègre, *The Behavior of the Earth: Continental and Seafloor Mobility* (Cambridge, MA: Harvard University Press, 1988).

153 **In his 1961 paper:** Robert S. Dietz, "Continent and Ocean Basin Evolution By Spreading of the Sea Floor," *Nature*, 190 (1961) 855–856.

153 **It was to be included:** Alan O. Allwardt, "The Roles of Arthur Holmes and Harry Hess in the Development of Modern Global Tectonics." Doctoral thesis, University of California, Santa Cruz, 1990, 192.

153 "**The marine geologists**": Menard, *The Ocean of Truth*, 154–55.

154 "**In late 1960, I initially**": Dietz, "Earth, Sea, and Sky; Life and Times of a Journeyman Geologist," 11–12.

155 **Hess wrote to the editor:** Allwardt, "The Roles of Arthur Holmes and Harry Hess in the Development of Modern Global Tectonics," 193.

155 **and the paper, retitled:** H. H. Hess, "History of Ocean Basins," in *Petrologic Studies—A Volume in Honor of A. F. Buddington* (1962).

156 "**Only one person**": Menard, The Ocean of Truth, 158.

157 "**The writer's attention**": Dietz, ed., *Ocean-Basin Evolution by Sea-Floor Spreading*, 12.

158 **"It was Dietz"**: Oreskes, *The Rejection of Continental Drift*, 271.

159 **"A principal objection"**: Dietz, "Continent and Ocean Basin Evolution by Spreading of the Sea Floor,"

159 **"A more acceptable mechanism"**: Hess, "History of Ocean Basins,"

159 **"To sum up"**: Holmes, *Principles of Physical Geology.*

160 **"In November 1960"**: Allwardt, "The Roles of Arthur Holmes and Harry Hess in the Development of Modern Global Tectonics," 194.

160 **Arthur Meyerhoff . . . published:** Arthur A. Meyerhoff, "Arthur Holmes: Originator of Spreading Ocean Floor Hypothesis," *J. Geophysical Research*, 73, no. 20 (1968).

161 **Both Dietz and Hess:** Robert S. Dietz, "Reply," J. Geophysical Research, 73, no. 20 (1968); H. H. Hess, "Reply," *J. Geophysical Res.*, 73, no. 20 (1968).

Chapter 12

165 **"One of the great obstacles"**: Daniel J. Boorstin, *The Discoverers* (New York: Random House, 1983), 86.

165 **Until the last two decades:** D. E. Wilhelms, *To a Rocky Moon: A Geologist's History of Lunar Exploration* (Tucson: University of Arizona Press, 1993), 352.

166 **They reported astounding:** Jack Green, "Hookes and Spurrs in Selenology," in *Geological Problems in Lunar Research*, ed. H. E. Whipple (New York: New York Academy of Sciences, 1965), 375.

168 **"Meteoritic impact"**: Jack Green, "Lunar Defluidization and Volcanism," in ibid., 455.

170 **In his Micrographia:** W. G. Hoyt, *Coon Mountain Controversies: Meteor Crater and the Development of Impact Theory* (Tucson: University of Arizona Press, 1987), 9–10.

170 **The accepted author:** Ibid., 11.

170 **"sheep, pygmy zebra"**: Ibid.

171 **He maintained this position:** Ibid., 14.

171 **"presents to the telescope"**: Ibid., 30.

171 **"Surely one of the"**: Wilhelms, *To a Rocky Moon*, 7.

171 **"closest thing to a saint"**: Hoyt, *Coon Mountain Controversies,* 37.

171 **Whereas most in the audience:** Ibid., 31.

172 **He decided to visit:** Ibid., 32.

172 "**The man who can**": Ibid., 38.

172 **"no conclusion" is:** Ibid., 39.

172 "**star entered the hole**": Ibid., 41.

173 **"the presumption was in favor . . .":** G. K. Gilbert, "The Origin of Hypotheses, Illustrated by Discussion of a Topographic Problem," *Science*, N. S. III, no. 53, p. 6 (1896).

173 **Gilbert delivered the paper:** Ibid., 49.

174 "**like plums**": Ibid., 52.

174 "**So useless has**": Wilhelms, *To a Rocky Moon*, 8.

174 "**The volcanic theory**": Hoyt, *Coon Mountain Controversies,* 59.

175 "**The lunar craters**": Ibid., 61.

176 **On a warm Tucson night:** Ibid., 73.

176 **On his return home:** Ibid., 74.

177 **Gilbert had not had:** Ibid., 252.

177 "**Surely it is a new**": Ibid., 253.

177 **Opik would become:** Wilhelms, *To a Rocky Moon*, 12.

178 **"The impacts made by . . .":** Hoyt, *Coon Mountain Controversies*, 259.

178 **If more than 500,000:** . .": Ibid., 264.

179 "**which is most rarely**": Hoyt, *Coon Mountain Controversies*, 203.

179 "**The contradiction is so**": See Ellen T. Drake, and others, "Origin of Impact Craters; Ideas and Experiments of Hooke, Gilbert, and Wegener," *Geology*, 12, no. 7 (1984).

179 **He proposed:** Dietz, "Earth, Sea, and Sky; Life and Times of a Journeyman Geologist," 21.

179 "**The disrupted nest**": Ibid.

180 **The archetypal outsider:** Ibid.

180 "**cosmic rather than volcanic**": Ibid., 21.

180 **He wrote up his thoughts:** Ibid.

181 **The Journal published:** Ibid.

Chapter 13

183 **In research, self-satisfaction:** Jacques Monod, Interview in *Nouvel Observateur* (1965).

183 "**long valleys**": Ralph B. Baldwin, "An Overview of Impact Cratering," *Meteoritics*, 13 (1978), 368.

183 "**the nearly tangential**": Ibid., 369.

184 "**the circular maria**": Ibid.

184 "**knowing far more**" Ibid.

184 "**nobody else**" . . . "**a significant field**": Ibid., 370.

184 **After spending the wartime years:** Ibid.

184 **"The only reasonable interpretation":** Ralph B. Baldwin, *The Face of the Moon* (Chicago: University of Chicago, 1949), 135.

185 "**To claim that**": Ibid., 146.

185 **To his credit:** Wilhelms, *To a Rocky Moon,* 20.

185 "**If anything**": Stephen G. Brush, "Nickel for Your Thoughts: Urey and the Origin of the Moon," *Science*, 217 (1982), 897.

185 **Urey read:** Ibid., 892.

186 **Urey published his ideas:** Brush, "Nickel for Your Thoughts: Urey and the Origin of the Moon," 891.

186 **Shoemaker's remarkable career:** See David Levy, Shoemaker by Levy (Princeton: Princeton University Press, 2000).

186 "**Why, we're going to**": Wilhelms, *To a Rocky Moon*, 20.

187 **Shoemaker won them:** Ibid., 21.

188 **his last paper:** Walter Bucher, "The Largest So-Called Meteorite Scars in Three Continents as Demonstrably Tied to Major Terrestrial Structures," published in *Geological Problems in Lunar Research*, ed. Whipple.

188 **just before he died:** Ursula B. Marvin, "Impact and Its Revolutionary Implications for Geology," in *Global Catastrophes in Earth History*, ed. Virgil L. Sharpton and Peter D. Ward (Boulder, Co: Geological Society of America, 1990), 152.

188 **In his 1949 manifesto:** Baldwin, *The Face of the Moon,* 91.

189 "**How many other gigantic**": Ibid., 92.

189 **In answer to:** Wilhelms, *To a Rocky Moon*, 62.

189 **In his opening paper:** Jack Green, "Josiah Edward Spurr," *Proc. Geol. Soc. Amer.*, 1968 (1971).

189 **In four volumes published:** Wilhelms, *To a Rocky Moon*, 13.

189 **Spurr's lunar grid:** Ibid.

189 **Green published a paper:** Jack Green, "Copernicus as a Lunar Caldera," *J. Geophysical Res.*, 76, no. 23 (1971).

190 "**Cost considerable time**": Wilhelms, *To a Rocky Moon*, 27.

Chapter 14

191 "**I can state flatly**": Lord Kelvin, Website (available from http://zapatopi.net/kelvquote.html.)

191 **When Shoemaker heard:** Wilhelms, *To a Rocky Moon* 27.

192 **One of its most important:** Ibid.

192 **In 1960, Don Gault:** Ibid., 48.

193 **Although most astronomers:** Ibid., 354.

195 "**even the antenna**": Ibid., 138.

195 "**It's basalt, isn't it?**": Ibid., 143.

195 "**Mother Nature**": Ibid.

196 "**Surveyor had to land**": Ibid., 153.

197 **When scientists opened:** Stephen G. Brush, *A History of Modern Selenogony: Theoretical Origins of the Moon, from Capture to Crash 1955–1984* (Dordecht: Kluwer Academic Press, 1988), 230.

198 **The analysis of samples:** Wilhelms, *To a Rocky Moon*, 212. See also Bevan M. French, *The Moon Book* (Harmondsworth, New York: Penguin Books, 1977).

Chapter 15

201 "**My mother groan'd**": William Blake, "Infant Sorrow," in *Songs of Experience* (1794).

202 **"In 1911":** Ursula B. Marvin, *Continental Drift: Evolution of a Concept* (Washington, DC: Smithsonian Institution, 1973), 65.

203 **"More than four decades later":** Howard B. Baker, The Earth Participates in the Evolution of the Solar System (Occasional Papers, Detroit Academy of Sciences, 1954), 16.

203 **"The separating force":** Ibid., 17.

203 **"No one seems to have paid any attention":** Stephen G. Brush, "Theories of the Origin of the Solar System 1956–1985," *Reviews of Modern Physics,* 62, no. 1 (1990), 85.

203 **Calculations by a colleague:** Ibid., 85.

204 **Menzel was favorably impressed:** Brush, *A History of Modern Selenogony* 247.

204 **Baker submitted a manuscript:** Ibid.

204 **The editor did:** Ibid.

204 **But most seem:** Ibid.

204 **Even the giant impactor:** Ibid., 248.

205 "**the probability of the planet**": Ibid., 248.

205 "**hit the earth**": H. F. S. Cooper, "Letter from the Space Center," *The New Yorker (*June 1987), 79.

206 "**A large, imposing man**": Ibid.

206 "**We're working on**": Ibid.

207 "**Cut the Gordian knot**": Ibid., 80.

207 "**No one had paid**": Ibid.

208 "**When I went to**": Ibid.

208 **Hertmann countered:** Ibid.

208 "**My point was**": Ibid.

209 "**breathed new life**": R. A. Kerr, "Making the Moon from a Big Splash," *Science*, 226 (1984), 1060.

209 "**A lot more work**": D. J. Stevenson, "Origin of the Moon—The Collision Hypothesis," *Ann. Rev. Earth Planet. Sci.*.

211 **Are we here:** Stuart Ross Taylor, *Destiny or Chance: Our Solar System and Its Place in the Cosmos* (Cambridge & New York: Cambridge University Press, 1998).

Chapter 16

213 "**What is this talked-of**" Robert Frost, "Riders," in *The Poetry of Robert Frost.,* ed. Edward Lathem, (New York: Holt, Reinhart & Winston 1969), p. 267.

214 "**The impact of solid bodies**": Eugene M. Shoemaker, "Why Study Impact Craters?" in *Impact and Explosion Cratering*, ed. D. Roddy, R. O. Pepin, and R. B. Merrill (New York: Pergamon Press, 1977), p. 1.

217 "**if microbes existed**": Curt Mileikowsky, and others, "Natural Transfer of Viable Microbes in *Space*," Icarus, 45 (2000), 391–392

218 **In the last 300:** Taylor, *Destiny or Chance*, 177.

Chapter 17

223 "**The moving finger**": Edward FitzGerald, trans., "The Rubáiyát of Omar Khayyám" (1859).

Acknowledgments

I AM GRATEFUL TO Brent Dalrymple, to Oberlin College geology professors William Skinner and Bruce Simonson, and to Gregory Davis of the University of Southern California for their comments on portions of the manuscript. My agent, John Thornton, is a true professional. My editor, Stephen Morrow, made many helpful suggestions.

The reader interested in studying the discoveries of twentieth-century geology in greater depth will find, as I have done, outstanding sources. Dalrymple's *The Age of the Earth* is a masterpiece. *The Rejection of Continental Drift,* by Naomi Oreskes, is essential reading for anyone interested in the history of geology and science. William Glen's *The Road to Jaramillo* chronicles the paleomagnetic underpinning of the plate tectonic revolution as well as it could possibly be done. *The Ocean of Truth,* William Menard's firsthand account of the post–World War II years in marine geology, is a precious legacy. Don Wilhelm's *To a Rocky Moon* tells the story of Apollo in masterful fashion. *Coon Mountain Controversies,* William Hoyt's history of Meteor Crater, could not be beat. *Shoemaker by Levy* provides a much-needed biography of the one I call the leading geologist of the second half of the twentieth century. My debt to these authors will be immediately and continuously apparent.

I dedicate this book to Joan, my *sine qua non.*

Index

Index

Index

Index

Index

Index

Index

Index